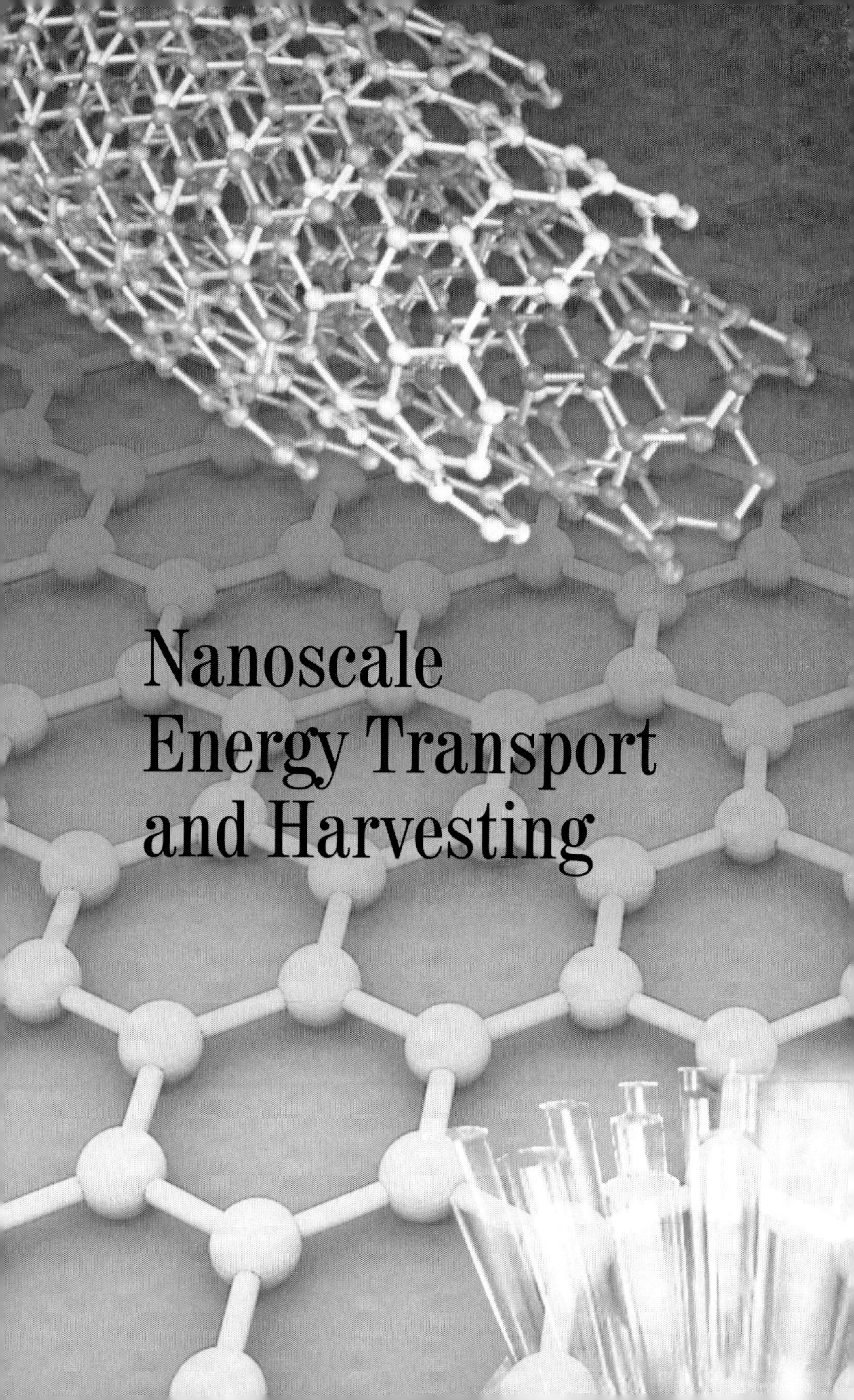

Nanoscale Energy Transport and Harvesting

Nanoscale Energy Transport and Harvesting

A COMPUTATIONAL STUDY

edited by

Gang Zhang

PAN STANFORD PUBLISHING

Published by

Pan Stanford Publishing Pte. Ltd.
Penthouse Level, Suntec Tower 3
8 Temasek Boulevard
Singapore 038988

Email: editorial@panstanford.com
Web: www.panstanford.com

British Library Cataloguing-in-Publication Data
A catalogue record for this book is available from the British Library.

Nanoscale Energy Transport and Harvesting: A Computational Study

ISBN 978-981-4463-02-7 (Hardcover)
ISBN 978-981-4463-03-4 (eBook)

Printed in the USA

Contents

Preface

Energy shortage is a great bottleneck in the supply of energy resources to an economy. The world's power demands are expected to rise 60% by 2030. Actually, people can solve the global energy crisis by enhancing the utilization efficiency of energy. Today, approximately 80% of the world's power is generated by heat engines that use fossil fuel combustion as a heat source, which is believed to be responsible for a large fraction of carbon dioxide emissions worldwide. The heat engines used in most thermal power station typically operate at 30–40% efficiency. This means that roughly 10 TW of heat energy is lost to the environment. Thermoelectric modules can potentially convert part of the wasted heat directly into electricity, reduce the usage of fossil fuels, and lower carbon emission. Moreover, microelectronic processors generate huge amount of heat in very small areas. Traditionally, this heat is considered as waste and may lead to the partial or total loss of the functionality of the processors. Power dissipation issues have recently become one of the greatest challenges for integrated electronic devices, and it is becoming a bottleneck for further development of smaller and faster devices. Currently, for every kilowatt-hour of energy consumed by a computer in a data centre, another kWh is needed for cooling. With the application of advanced thermal management and energy conversion technologies, world's household PCs can be converted to billions of mini power plants and up to 50% electric energy can be saved.

In addition, thermal management is also significantly important for solar energy harvesting. The solar cell technology can harvest and convert part of the solar energy into electricity by using the photovoltaic effect. The efficiency of conventional solar cells is usually quite low and limited because about 50% of the solar energy is lost to heat through radiation to the environment. Based on the thermoelectric effect, it is in principle possible to further convert the heat energy to electricity, which provides a new channel for solar energy harvest and may significantly improve the efficiency of solar cells.

Thermoelectric energy conversion efficiency depends on the figure of merit *ZT*, which is proportional to the Seebeck coefficient, electrical conductivity, and absolute temperature, but inversely proportional to thermal conductivity. Recently the basic possibility of significantly increasing *ZT* through creation of new classes of thermoelectric materials with low-dimensional nanoscale structures has been demonstrated. The term nanoscale systems denote structures composed of a limited, small number of atoms. The interest of the scientific community in nanoscale systems has been boosted by the recent advent of micromanipulation techniques and nanotechnologies. Nanoscale materials have generated broad excitement both for fundamental science and for their potential applications in technology because of their scientific richness and promise in technological applications involving various devices. Efficient conversion between different forms of energy: thermal, electric, and optical, is a key enabler in many areas of science and engineering. Development of nanofabrication, characterization, measurement, and atomistic simulation tools can contribute to inspire new and better technologies for potential applications in energy saving and conversion.

The application of nano energy devices has highlighted the need for greater quantitative understanding of materials at the nanoscale. Understanding the physics of such systems by computational study is particularly important because their small size makes it is challenging to apply standard experimental measurement methods. Over the past decades, advances in computer science have spurred advances in fundamental theoretical techniques, mathematical modelling, and numerical simulation, giving rise to a revolution with extraordinary impact on nanoscience and nanoengineering.

The aim of this book is to provide an introduction for both theorists and experimentalists to the current computational technology and then looking at the applications of nanostructures in renewable energy and the associated research topics. The book should also be useful for graduate-level students who want to explore this new field of research. The book addresses the current and commonly used computational technologies and their applications in study of nanoscale energy transport and conversion. With content relevant to both academic and commercial viewpoints, the book will interest researchers and postgraduates as well as consultants in the renewable energy industry.

The chapters have been written by internationally recognized experts in computational physics and provide in-depth introductions to the directions of their research. This approach of a multiauthor reference book appeared to be particularly useful in view of the vast amount of literature available on different forms of computational study. While there exists excellent reviews highlighting single facets of computational methods for renewable energy, we feel that the field lacks a reference that brings together the most important contributions to this topic in a comprehensive manner. This book is an attempt to fill the gap. Along these lines, our intention was to embed research on new energy materials into a wider context of computational researches. We thus hope that this book may serve as a catalyst both to fuse existing computational approaches and to inspire new computational tools in the rapidly growing area of new energy material research.

Accordingly, the book is organized into five chapters: The first chapter features a pedagogical introduction to molecular dynamics simulations. For large systems, molecular dynamics is a useful tool for investigating atomic motion. The trajectories of molecules and atoms can be determined empirically using a force field. The applications of molecular dynamics simulation have covered a wide range of research topics, such as liquids, defects, fatigue, surface, clusters, and biomolecules. Therefore, molecular dynamics simulation has become indispensable in today's research of physical and material science. The second chapter outlines the ballistic phonon transport theory in a quasi-one-dimensional (quasi-1D) system whose length is much shorter than the coherence length, which is bound by phonon scattering events. Landauer approach for describing the coherent phonon transport in a quasi-1D system was introduced. Thermal conductance of carbon nanotubes is reviewed.

Chapter three systematically discusses the non-equilibrium Green's function (NEGF) method and tries to offer a complete theory for the investigation of quantum thermal transport. The NEGF method is widely and successfully used for the study of electronic transport. Here the method is generalized for thermal transport. After this theoretical opening, thermal conductance of graphene, a two-dimensional crystal consisting of a single atomic layer of carbon, has been reviewed. The following chapter summarizes groundbreaking work on ballistic phonons thermal transport at low temperature in low-dimensional quantum structures. Within the Landauer transport theory, the authors present a general formula to calculate the ballistic thermal conductance associated with phonon in the linear response limit. Then, a comparative analysis for the ballistic thermal transport is made between two-dimensional and three-dimensional models. The fifth chapter provides a systematic review on the effect of surface passivation on the phonon thermal conductivity and thermoelectric property of nanowires, based on introduction of molecular dynamics simulations. The underlying physical mechanism and analysis method are also presented. These results are helpful to understand the enhancement of thermoelectric performance of nanomaterials and to the design of renewable energy devices.

I finally remark that the various points of view expressed in the single chapters may not always be in full agreement with each other. As editor, I do not necessarily aim to achieve a complete consensus among all authors, as differences in opinions are typical for a very active field of research such as the one presented in this book.

I am most grateful to Stanford Chong, director of Pan Stanford Publishing, for his invitation to edit this book, and for his help in getting the project started. We also thank Sarabjeet Garcha from Pan Stanford Publishing for kind and efficient assistance in editing this book. I gratefully acknowledge financial support from National Natural Science Foundation of China (Grant No. 11274011) and the Ministry of Education of China (Grant No. 20110001120133).

I finally wish to thank all book chapter authors for sharing their expertise in this review volume. Their strong efforts and enthusiasm for this project were indispensable for bringing it to success.

Gang Zhang
Beijing
Winter 2014

Chapter 1

Molecular Dynamics Simulations for Computing Thermal Conductivity of Nanomaterials

Jie Chen,[a] Gang Zhang,[b] and Baowen Li[a,c]

[a]*Department of Physics, Centre for Computational Science and Engineering, and Graphene Research Centre, National University of Singapore, Singapore 117542, Singapore*

[b]*Institute of High Performance Computing, Singapore 138632, Singapore*

[c]*NUS-Tongji Center for Phononics and Thermal Energy Science and School of Physical Science and Engineering, Tongji University, Shanghai 200092, People's Republic of China*

zhangg@ihpc.a-star.edu.sg

1.1 Introduction to Molecular Dynamics

The physical properties of matter are found in its structure and motion of its constituent building blocks, and the dynamics are contained in the solution to the many-body problem. The many-body problem originated from the dynamics of the solar system, and its analytic solution turns out to be insoluble for three or more bodies. Although it is quantum mechanics instead of classical mechanics that describes the fundamental physics of condensed matter, the

Nanoscale Energy Transport and Harvesting: A Computational Study
Edited by Gang Zhang

ISBN 978-981-4463-02-7 (Hardcover), 978-981-4463-03-4 (eBook)
www.panstanford.com

attempt to numerically solve the Schrödinger equation for a system of many nuclei and electrons is too formidable and not at all feasible in practice. Thus, one has to resort to approximations.

Molecular dynamics (MD) simulation is an extremely powerful tool to handle many-body problems at atomic level based on classical mechanics, which numerically solves Newton's equation of motion for a many-body system. It has the advantage of simulating realistic material with accurate many-body interatomic interaction obtained from first-principles calculations, which was not available but simplified by the two-body potentials with analytical form in the earlier theoretical model. The applications of MD simulation have covered a wide range of research topics, such as liquids [1], defects [2], fatigue [3], surface [4, 5], clusters [6, 7], and biomolecules [8]. Therefore, MD simulation has become indispensable in today's research of physical and material science.

The typical feature size that current first-principle calculations, such as density-functional theory (DFT), can be used to explore the thermal properties of nanostructures is on the order of several nm [9–11]. With MD simulations, the system size under study can be enlarged a lot. For instance, MD simulations of silicon nanowires with length up to μm [12] and cross-sectional area up to 806 nm^2 [13] have been reported. Moreover, Markussen *et al.* [9] studied thermal properties of thin silicon nanowires with both DFT and classical calculations based on Tersoff potential. They found that the calculation results of thermal conductance obtained from DFT and Tersoff calculations agree within 10% [9].

The validity of the classical approximation can be evaluated based on the de Broglie thermal wavelength [14] defined as

$$\Lambda = \sqrt{\frac{2\pi\hbar^2}{mk_B T}}, \tag{1.1}$$

where $\hbar$ is Planck's constant, m is the atom mass, k_B is Boltzmann's constant, and T is the temperature. The classical approximation is valid if $\Lambda << a$, where a is the nearest neighbor separation. Under this condition, the entire system can be treated as dilute gas model based on which the classical kinetic gas theory is formulated [15]. In this case, each molecule can be considered as a classical particle with a well-defined position and momentum. Moreover, different molecules are distinguishable from each other [15]. Take the diamond-structured silicon for instance, its lattice constant is $L_0 = 5.43$ Å, and $a = \sqrt{3}L_0/4 = 2.35$ Å. The de Broglie thermal wavelength at room temperature is $\Lambda = 0.19$ Å, which is one order of magnitude smaller than the nearest neighbor separation. Therefore, for the group IV semiconductors with diamond structure, such as silicon and germanium, the classical approximation is appropriate above room temperature from the dilute gas point of view. Furthermore, quantum effects become important for any system when the temperature is sufficiently low. The drop of the heat capacity for crystals below the Debye temperature is a well know example of measurable quantum effects in solids. This is of particular importance in the study of thermal transport [16].

In the context of thermal transport, MD simulation is a statistical mechanics approach, which relates the microscopic behavior in a system with its thermodynamics. According to statistical mechanics, physical quantities can be evaluated by averaging over configurations distributed according to a certain statistical ensemble. For the N-particle system, MD simulation calculates the trajectory in a $6N$-dimensional phase space ($3N$ positions and $3N$ momenta) at each instantaneous time. This trajectory obtained from MD simulation provides such a set of configurations. To get the thermodynamic variables such as temperature, it relies on the ergodicity hypothesis of statistical mechanics, which asserts that the phase space can be fully recovered in the long-time limit and the time average is equivalent to the ensemble average. Therefore, thermodynamic variables of interest can be obtained by the time average along the trajectory in MD simulation.

A typical procedure of MD simulation can be described by the following key steps. To begin with, an appropriate force field that can accurately describe the interatomic interaction in the system of interest is required. The next step is to construct the initial position and velocity of each atom in the system, and generate a list for the nearest neighbor interaction. Usually atoms are initially located at their equilibrium positions and assigned with random velocity according to Gaussian distribution. After that, one calculates the force applied to each atom according to the force field, numerically solves Newton's equation of motion, and updates the position and velocity of each atom. This is a major step in MD simulation, which involves many technical details such as boundary conditions, numerical integration algorithm, and heat bath (reservoir) that can realize the canonical ensemble. Finally, one calculates the ensemble average of the thermodynamic variables of interest with long enough simulation time.

1.2 Force Field Potential

The interatomic force in MD simulation is derived from the system's potential energy. Although a fully quantum-mechanical treatment of the system's Hamiltonian is highly desirable, it can only be applied to a system with very small size due to the large computational load. Therefore, tremendous efforts have been put in the past to develop reliable empirical force field in many material systems that can accurately describe the interatomic interactions and easily be applied to a large system size. The simplest model for interatomic interaction is the van der Waals (vdW) force with pairwise (two-body) interaction. For the covalent bonding, more complicated form of force fields with many-body interaction are used. To illustrate this in more details, we introduce several widely used force field potentials as examples.

1.2.1 *Pair Potential*

The Lennard–Jones (LJ) potential is a widely used pair potential to model the vdW force. It is defined as

$$V(r) = 4\varepsilon\left[\left(\frac{\sigma}{r}\right)^{12} - \left(\frac{\sigma}{r}\right)^{6}\right], \tag{1.2}$$

where ε and σ are characteristic energy and length unit, respectively, and r is the interatomic distance. It is also known as 12-6 potential and can be expressed as

$$V(r) = \frac{A}{r^{12}} - \frac{B}{r^{6}}, \tag{1.3}$$

with A and B denoting the repulsive and attractive constant in LJ potential. As shown in Fig. 1.1, the equilibrium distance in LJ

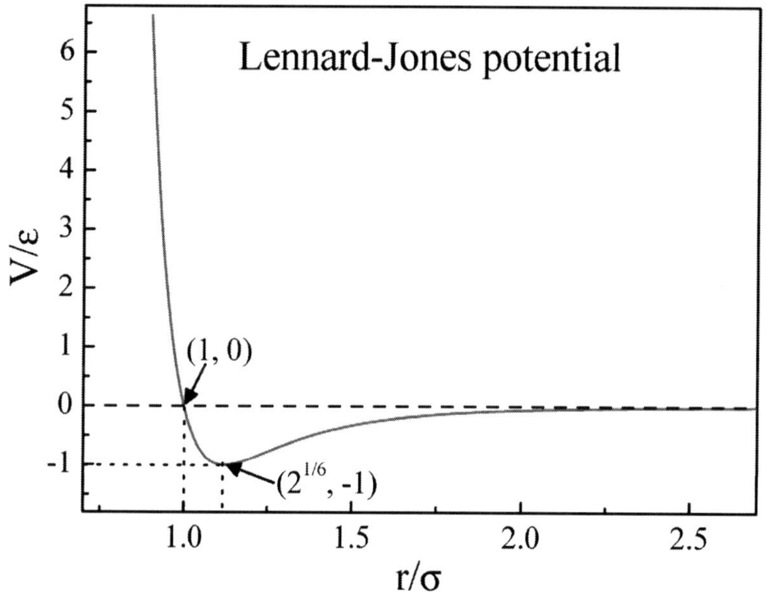

Figure 1.1 Schematic picture for Lennard–Jones potential.

potential is at

$$r_0 = 2^{1/6}\sigma = \left(\frac{2A}{B}\right)^{1/6}, \tag{1.4}$$

and the well depth is

$$\varepsilon = \frac{B^2}{4A}. \tag{1.5}$$

Moreover, LJ potential vanishes when the interatomic distance is greater than 2.5σ, which can be used as the cut-off distance in practical simulations to reduce the computer time for force field calculations.

LJ potential is often used to describe the properties of gases, simple fluids, and is particularly accurate for noble gas atoms. It is not able to account for the directional nature or the environmental dependence of the bond in many solid state systems where many-body effect is important. Therefore, LJ potential is not suitable to model systems with practical interest, such as semiconductors. For such systems a more complex many-body potential should be

Table 1.1 Parameters in Lennard–Jones potential for different systems

	σ (Å)	ε (meV)
[a]He	3.25	1.72
[a]Ne	2.93	1.12
[a]Ar	3.37	10.34
[a]Kr	3.66	15.51
[a]Xe	4.05	21.80
[b]Graphene	3.41	2.39
[b]C_{60}	3.47	2.86
[b]Graphene-C_{60}	3.44	2.62
[c]C-Si	3.63	8.91
[c]C-O	3.27	3.44
[c]C-B	3.53	5.96
[c]C-N	3.35	3.69

[a]From Ref. [18a].
[b]From Ref. [18b].
[c]From Ref. [19] based on mixing rule (see Section 1.2.3).

used, which takes into account the many-body interaction. However, regardless of how well it is able to model actual materials, the importance of LJ potential cannot be underestimated as it represents an important model to describe the vdW interaction. LJ potential is the standard potential to use when the investigation is focused on the fundamental physics, rather than studying the properties of a specific material. For instance, LJ potential is widely used in literatures to model graphene-substrate interaction and inter-layer interaction in multi-layer graphene [17]. Table 1.1 summarizes the LJ potential parameters for different systems recorded in literatures.

1.2.2 *Many-Body Potential*

The Stillinger–Weber (SW) potential [20] is a popular many-body potential to model the crystalline silicon. It consists of both two-body and three-body terms defined as

$$v_2(\mathbf{r}_i, \mathbf{r}_j) = \varepsilon f_2(r_{ij}/\sigma), \tag{1.6}$$

$$v_3(\mathbf{r}_i, \mathbf{r}_j, \mathbf{r}_k) = \varepsilon f_3(\mathbf{r}_i/\sigma, \mathbf{r}_j/\sigma, \mathbf{r}_k/\sigma), \tag{1.7}$$

where $\mathbf{r}_i$ is the position of atom i, r_{ij} is the distance between atom i and j, and ε and σ denote the energy and length unit, respectively.

These two units are determined in such a way that ε is chosen to give f_2 depth-1, and σ is chosen to make $f_2(2^{1/6})$ vanish.

The reduced pair potential f_2 reads

$$f_2(r) = \begin{cases} A(Br^{-p} - r^{-q})\exp[(r-a)^{-1}], r < a \\ 0, r \geq a \end{cases}, \tag{1.8}$$

where a is the cut-off distance beyond which interaction vanishes. The three-body interaction f_3 is defined as

$$f_3(\mathbf{r}_i, \mathbf{r}_j, \mathbf{r}_k) = h(r_{ij}, r_{ik}, \theta_{jik}) + h(r_{ji}, r_{jk}, \theta_{ijk}) + h(r_{ki}, r_{kj}, \theta_{ikj}), \tag{1.9}$$

$$h(r_{ij}, r_{ik}, \theta_{jik}) = \lambda \exp[\gamma(r_{ij}-a)^{-1} + \gamma(r_{ik}-a)^{-1}]\left(\cos\theta_{jik} + \frac{1}{3}\right)^2, \tag{1.10}$$

where θ_{jik} is the angle between $\mathbf{r}_j$ and $\mathbf{r}_k$ subtended at vertex i. The ideal tetrahedral geometry of the diamond structure is favored by the angle $\theta_t = \cos^{-1}(-1/3)$ in Eq. 1.10.

Although SW potential can model the properties of diamond-structured silicon quite well, it fails to describe the behavior of non-tetrahedral polytypes of silicon. To overcome this limitation, Tersoff proposed a new approach [21] to model covalent interaction based on the concept of bond order: the strength of a bond between two atoms depends on the local environment. It is a summation of two-body interactions expressed as [22]

$$E = \frac{1}{2}\sum_{i \neq j} V_{ij}, \; V_{ij} = f_C(r_{ij})\left[f_R(r_{ij}) + b_{ij} f_A(r_{ij})\right], \tag{1.11}$$

where f_R and f_A describe the repulsive and attractive forces, respectively, and f_C is the cut-off function. The bond order term b_{ij} is defined as [22]

$$\begin{aligned} b_{ij} &= \chi_{ij}(1 + \beta_i^{n_i}\zeta_{ij}^{n_i})^{-1/2n_i}, \\ \zeta_{ij} &= \sum_{k \neq i, j} f_C(r_{ik})\omega_{ik} g(\theta_{ijk}), \\ g(\theta_{ijk}) &= 1 + c_i^2/d_i^2 - c_i^2/\left[d_i^2 + (h_i - \cos\theta_{ijk})^2\right], \end{aligned} \tag{1.12}$$

where the local information about another atom k is involved.

Based on Tersoff's formalism, Brenner included additional terms to account for nonlocal effect and extended the application to the hydrocarbon systems [23], which allows for bond breaking and creation with appropriate changes in atomic hybridization. The Brenner potential is later improved to give a better description of physical properties for hydrocarbon molecules [24].

1.2.3 *Mixing Rule*

The force field parameters are usually developed for individual element only. In addition to single element materials, alloys offer a continuously variable system with a wide range of crystal lattices and band gaps, leading to various physical and chemical properties. For instance, Silicon and Germanium can form a continuous series of substitutional solid. The continuous variation in the energy band gap in $Si_{1-x}Ge_x$ provides challenging opportunities for Si optoelectronic because it potentially allows emission and detection of the photons in large wavelength range. Recently, single crystalline $Si_{1-x}Ge_x$ NWs have been successfully grown. The experimental synthesis offers intriguing opportunities for the development of $Si_{1-x}Ge_x$ based nano thermoelectric devices. To model the interatomic interaction in multi-component system, the force field parameters are approximated by the combination of single-element parameters. The most frequently used mixing rules are the arithmetic average (Lorentz rule) for length parameter

$$\sigma_{ij} = \frac{\sigma_i + \sigma_j}{2}, \tag{1.13}$$

and the geometric average (Berthelot rule) for energy parameter

$$\varepsilon_{ij} = \sqrt{\varepsilon_i \varepsilon_j}. \tag{1.14}$$

1.3 Integration of the Equations of Motion

Molecular dynamics simulation specifies how the positions and velocities of the particles in the system vary with time. This can be generated by integrating Newton's laws of motion. In the simulation, the force on each particle will change whenever the particle and/or its neighbors change their positions. Under the influence of a realistic potential, the motions of all the particles are coupled together, giving rise to a many-body problem. There are many computational algorithms to integrate Newton's equations of motion numerically, with different levels of accuracy and complexity. The basic idea is that the integration is divided into many small stages, each separated in time by a fixed time step. The total force on each particle at a given time, t, is calculated based on the atomistic potential. From the calculation of force we can determine the accelerations of the particles, which are then combined with the positions and velocities at time t to calculate the positions and velocities at time $t + \Delta t$. In this time step, it is assumed that the force is a constant. At time $t + \Delta t$, based on the new positions of all particles, the forces on the particles are determined, leading to new positions and velocities at time $t + 2\Delta t$, and so on.

Here we introduce two popularly used algorithms: fourth-order Runge–Kutta method and velocity Verlet method.

Let $\mathbf{r}(t)$, $\mathbf{v}(t)$, and $\mathbf{a}(t)$ denote the position, velocity, and acceleration at time t, respectively. $V[\mathbf{r}(t)]$ is the potential energy, m is the mass, and Δt is the time step. To update these variables at time $t+\Delta t$, the fourth-order Runge–Kutta method is carried out based on the four-step evaluation as

$$\mathbf{q}_1 = \mathbf{v}(t)\Delta t, \quad \mathbf{p}_1 = \mathbf{a}(t)\Delta t, \tag{1.15}$$

$$\mathbf{q}_2 = \left(\mathbf{v}(t) + \frac{\mathbf{p}_1}{2}\right)\Delta t, \quad \mathbf{p}_2 = -\frac{V\left(\mathbf{r}(t) + \frac{\mathbf{q}_1}{2}\right)}{m}\Delta t, \tag{1.16}$$

$$\mathbf{q}_3 = \left(\mathbf{v}(t) + \frac{\mathbf{p}_2}{2}\right)\Delta t, \quad \mathbf{p}_3 = -\frac{V\left(\mathbf{r}(t) + \frac{\mathbf{q}_2}{2}\right)}{m}\Delta t, \tag{1.17}$$

$$\mathbf{q}_4 = \left(\mathbf{v}(t) + \mathbf{p}_3\right)\Delta t, \quad \mathbf{p}_4 = -\frac{V\left(\mathbf{r}(t) + \mathbf{q}_3\right)}{m}\Delta t, \tag{1.18}$$

$$\begin{aligned} \mathbf{r}(t+\Delta t) &= \mathbf{r}(t) + \frac{\mathbf{q}_1 + 2\mathbf{q}_2 + 2\mathbf{q}_3 + \mathbf{q}_4}{6}, \\ \mathbf{v}(t+\Delta t) &= \mathbf{v}(t) + \frac{\mathbf{p}_1 + 2\mathbf{p}_2 + 2\mathbf{p}_3 + \mathbf{p}_4}{6}, \\ \mathbf{a}(t+\Delta t) &= -\frac{V\left(\mathbf{r}(t+\Delta t)\right)}{m}, \end{aligned} \tag{1.19}$$

where $\mathbf{q}_n$ and $\mathbf{p}_n$ are auxiliary variables used in the evaluations.

The velocity Verlet method updates the position and velocity through the two-step evaluation as:

$$\mathbf{r}(t+\Delta t)=\mathbf{r}(t)+\mathbf{v}(t)\Delta t+\frac{1}{2}\mathbf{a}(t)\Delta t^2, \tag{1.20}$$

$$\mathbf{v}\left(t+\frac{\Delta t}{2}\right)=\mathbf{v}(t)+\frac{1}{2}\mathbf{a}(t)\Delta t, \tag{1.21}$$

$$\mathbf{a}(t+\Delta t)=-\frac{V\,(\mathbf{r}(t+\Delta t))}{m}, \tag{1.22}$$

$$\mathbf{v}(t+\Delta t)=\mathbf{v}\left(t+\frac{\Delta t}{2}\right)+\frac{1}{2}\mathbf{a}(t+\Delta t)\Delta t. \tag{1.23}$$

It is obvious that in any integration algorithm, the time step is always an important factor. Too large time step may arise in the integration algorithm due to high energy overlaps between atoms and lead to a program failure. Too small time step will result in much more computer time. Usually, in one time step, the motion of the particle should be approximately one order of magnitude smaller than the atom–atom bond length. For most solid state materials, the suggested time step at room temperature is from 0.1 fs to 1 fs (1 fs $= 10^{-15}$s).

1.4 Temperature in Molecular Dynamics

1.4.1 *Heat Bath*

Based on equipartition theorem, the temperature of a system in MD simulation is defined as

$$T=\left\langle\sum_{i=1}^{N} m_i\mathbf{v}_i\cdot\mathbf{v}_i\right\rangle/\left(3Nk_{\mathrm{B}}\right), \tag{1.24}$$

where N is total number of atoms in the system, k_{B} is Boltzmann's constant, m_i and $\mathbf{v}_i$ is the mass and velocity (vector form) of atom i, respectively, and the angular bracket denotes the ensemble average. To set the system at aimed temperature, heat bath is used to realize the canonical ensemble, that is, constant temperature. Depending on the nature of excitation, there are two kinds of heat baths widely used in MD simulation. One is the deterministic type, such as Nosé–Hoover (NH) heat bath [25, 26], and the other is the stochastic type, such as Langevin heat bath [27].

The evolution of the atoms in thermal contact with NH heat bath is ruled by the following equations

$$\frac{d\mathbf{q}_i}{dt} = \frac{\partial H}{\partial \mathbf{p}_i}, \ \frac{d\mathbf{p}_i}{dt} = -\frac{\partial H}{\partial \mathbf{q}_i} - \zeta \mathbf{p}_i, \tag{1.25}$$

where H is the Hamiltonian of the system, $\mathbf{p}_i$ and $\mathbf{q}_i$ are the momentum and coordinate of atom i, respectively, and ζ is an auxiliary variable modeling the microscopic action of the heat bath. The dynamics of ζ is governed by the following equation:

$$\frac{d\zeta}{dt} = \frac{1}{\tau^2}\left(\frac{\sum_{i=1}^{N}\frac{\mathbf{p}_i\cdot\mathbf{p}_i}{m_i}}{3Nk_\mathrm{B}T} - 1\right), \tag{1.26}$$

where T and τ are the aimed temperature and response time of heat bath, respectively, m_i is the mass of atom i, and N is the total number of atoms that are in contact with heat bath.

With Langevin heat bath, the equation of motion can be described as

$$\frac{d\mathbf{p}_i}{dt} = -\frac{\partial H}{\partial \mathbf{q}_i} + \xi - \lambda \mathbf{p}_i, \tag{1.27}$$

where the random force ξ and the dissipation rate λ are introduced into the system simultaneously. This kind of stochastic excitation is consistent with the microscopic picture of Brownian motion. The random force ξ follows the Gaussian distribution with zero mean value and variance of $2m\lambda k_\mathrm{B}T$ according to the fluctuation-dissipation theorem [28].

1.4.2 *Quantum Correction*

In a classical system, all the modes (degrees of freedom) are equally exited. Based on the equipartition theorem between the kinetic and potential energy, each mode has a constant energy of $k_\mathrm{B}T$, and thus the total energy of a classical system with N atoms is given by

$$E_\mathrm{C} = 3Nk_\mathrm{B}T_\mathrm{C}, \tag{1.28}$$

where T_C denotes the temperature of the classical system. The quantum-mechanical description of lattice energy E_Q is defined as

$$E_Q = \sum_{\lambda} \hbar\omega_{\lambda} \left(n_{\lambda} + \frac{1}{2} \right), \tag{1.29}$$

where ω_{λ} and n_{λ} are the eigenfrequency and occupation number for a given eigen-mode λ. When the system temperature is higher than the Debye temperature, the occupation number given by the Bose–Einstein distribution $n_{\lambda} = 1/\left(e^{\hbar\omega_{\lambda}/k_B T_Q} - 1\right)$ can be approximated by $k_B T_Q/\hbar\omega_{\lambda}$, and thus the associated energy for each mode is retarded to the classical case of $k_B T_Q$, which means all the modes contribute equally to the total energy (excluding the zero-point energy). However, at low temperature, there is a freezing out of high-frequency modes in the quantum system so that energy associated with different modes deviates from the classical case.

To qualitatively account for this discrepancy, a system-level mapping between the classical temperature used in MD simulation (T_{MD}) and the quantum temperature for the real environment (T_{real}) has been proposed by equating the total energy in the classical and quantum descriptions

$$3Nk_B T_{MD} = \int_0^{\omega_D} DOS(\omega) n(\omega, T_{real}) \hbar\omega d\omega, \tag{1.30}$$

$$3Nk_B T_{MD} = \int_0^{\omega_D} DOS(\omega) \left(\frac{1}{2} + n(\omega, T_{real}) \right) \hbar\omega d\omega, \tag{1.31}$$

where ω_D is the Debye frequency, $DOS(\omega)$ is the density of states for modes with frequency from ω to $\omega + \Delta\omega$. This equality permits a quantitative relation between T_{MD} and T_{real} based on the lattice dynamics information of the system, and both schemes with and without zero-point energy have been used in literature.

Figure 1.2 shows the typical relation between T_{MD} and T_{real} in SiNWs. Without zero-point energy (Eq. 1.30), the classical temperature T_{MD} can reach zero temperature and increases monotonically with the increase of quantum temperature T_{real}. Due to the inclusion of zero-point energy in Eq. 1.31, there exists nonzero minimum temperature for the classical temperature T_{MD}, which shifts up T_{MD} by a constant compared to that without zero-point energy (Fig. 1.2).

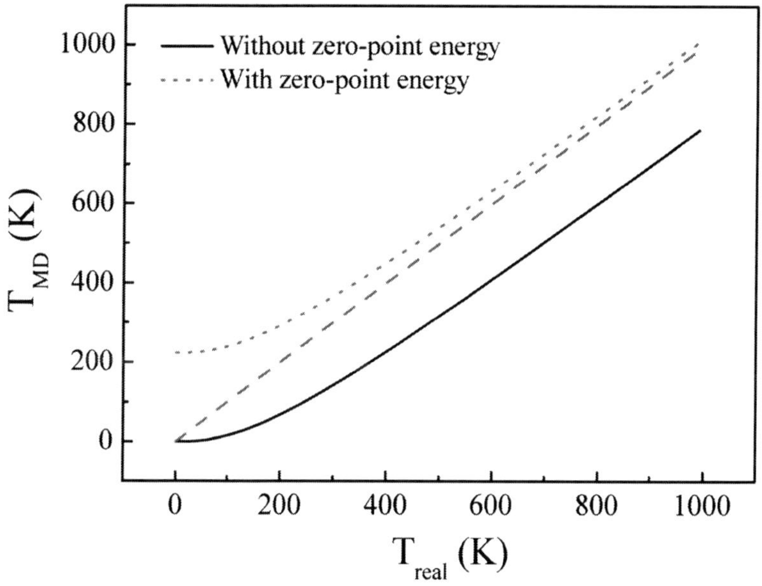

Figure 1.2 Classical temperature used in MD simulation (T_{MD}) versus quantum temperature for the real environment (T_{real}) in SiNWs. The solid and dotted line draws T_{MD} without and with zero-point energy, respectively. The dashed line draws T_{real} for reference.

As a result, the mapping between T_{MD} and T_{real} is not unique but depends on the exact definition of the system energy.

The quantum correction to the thermal conductivity is a post-processing step after MD simulation. By assuming the heat flux is equal in the classical and quantum systems

$$J = -\kappa_{\text{MD}} \nabla T_{\text{MD}} = -\kappa_{\text{real}} \nabla T_{\text{real}}, \tag{1.32}$$

thermal conductivity is rescaled by quantum correction as

$$\kappa_{\text{real}} = \kappa_{\text{MD}} \frac{\partial T_{\text{MD}}}{\partial T_{\text{real}}}, \tag{1.33}$$

with a scaling factor $\frac{\partial T_{\text{MD}}}{\partial T_{\text{real}}}$ which can be calculated from Eq. 1.30 or 1.31.

Figure 1.3 shows the typical quantum correction scaling factor in SiNWs derived from the temperature mapping shown in Fig. 1.2. With quantum correction, it can qualitatively take into account the decrease of heat capacity in crystalline solids at low-temperature by

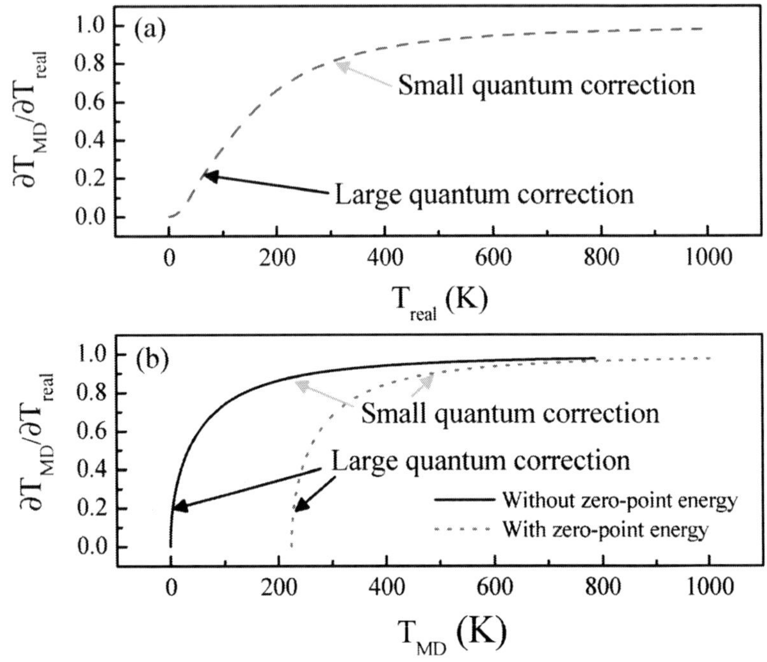

Figure 1.3 Quantum correction scaling factor ($\partial T_{MD}/\partial T_{real}$) in SiNWs versus (a) quantum temperature (T_{real}) and (b) classical temperature used in MD simulation (T_{MD}). The mapping between T_{MD} and T_{real} is shown in Fig. 1.2.

the decrease of scaling factor, while it results in a small quantum correction above Debye temperature where quantum effect are negligible (Debye temperature for Si is 645 K [29]). Therefore, this phenomenological quantum correction technique can at least *qualitatively* capture the temperature effect on thermal transport in crystalline solids. However, since this classical-quantum mapping is done based on the system-level properties (e.g., temperature), it cannot reproduce the actual quantum-mechanical microscopic properties (e.g., relaxation time and occupation number) at low temperature, leading to the discrepancy between MD simulation with quantum correction and fully quantum-mechanical approach in the prediction of thermal conductivity below Debye temperature [16]. Besides, although both quantum correction schemes (with and without zero-point energy) give rise to a small quantum correction in the high-temperature regime, it actually predicts different scaling factors for a given MD simulation temperature T_{MD} in the low-temperature regime (Fig. 1.3b), due to the different mapping between T_{MD} and T_{real} in different schemes (Fig. 1.2). Therefore, one should be aware of the limitation of this type of system-level quantum correction that it is only useful in a qualitative sense.

1.5 Non-equilibrium Molecular Dynamics

1.5.1 *Background*

There are two different approaches in non-equilibrium molecular dynamics (NEMD) simulations. One approach is to set up a temperature gradient across the system by fixing the two ends of the system at different temperature, and then calculate the resultant heat flux running through the system in the non-equilibrium steady state. In this approach, temperature of the two ends is fixed by putting the two ends in contact with heat bath, which induces additional force to the particles in it and the dynamics of the whole system is governed by the Hamiltonian. This approach is in analogy to real experimental measurement, and is also known as direction method. More importantly, it is consistent with the actual physical picture of heat conduction: the temperature gradient across the system is the "*cause*" of heat conduction, and consequently the heat flux running through the system is the "*effect*". Moreover, the way to fix temperature with heat bath, such as Langevin heat bath, is based on Brownian dynamics which is the well-known microscopic picture of the motions of the particles, and is also consistent with the fluctuation–dissipation theory [28]. The deficiency of this approach is that it usually requires simulation for a long period to obtain the constant heat flux in the non-equilibrium steady state.

The other approach is to reverse the usual cause and effect picture. The "effect", *heat flux*, is now imposed to the system, and the "cause", *temperature gradient*, is obtained from simulations. In this approach, the constant heat flux is artificially introduced to the two ends of the system, by either exchanging the velocity of the

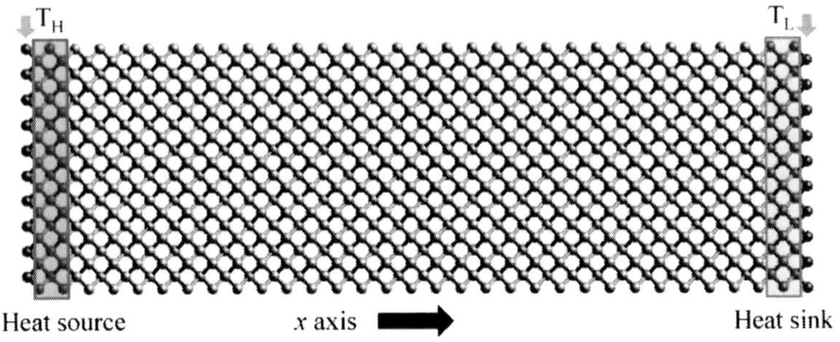

Figure 1.4 Schematic picture for NEMD simulation of SiNWs. The longitudinal direction is set along x-axis. The green arrows point the boundary atoms. Atoms in red and blue boxes are put in contact with the heat bath at high temperature T_H and low temperature T_L, respectively.

particles at two ends [30] or scaling the velocity of the particles at two ends [31]. This approach has the advantage that the slowly converging quantity *heat flux* is now known exactly and need not to be calculated. As a result, total simulation time can be reduced as only the relatively fast converging quantity *temperature* needs to be calculated in the simulation. However, the drawback of this approach is obvious: it is not consistent with the actual physical picture of heat conduction. Moreover, the heat flux is artificially introduced and there is no fundamental theoretical basis for it. As a result, the dynamics of the whole system is not completely governed by the Hamiltonian.

Therefore, the first approach in NEMD simulation is employed throughout this chapter. Figure 1.4 shows the schematic picture for NEMD simulation of silicon nanowires (SiNWs). Here we set longitudinal direction along x-axis, and atoms in the same layers means they have the same x-coordinate. Fixed boundary condition (BC) is imposed on the boundary layers (indicated by arrows in Fig. 1.4) at two ends of SiNWs and the surface atoms, either fixed or free BC. Next to the boundary layers, several layers of SiNWs are put in contact with the heat bath (inside the rectangular box in Fig. 1.4), which can reproduce the canonical ensemble (i.e., constant temperature). To simulate the heat source and sink in real experiment, two heat bathes at high temperature T_H and low temperature T_L are used.

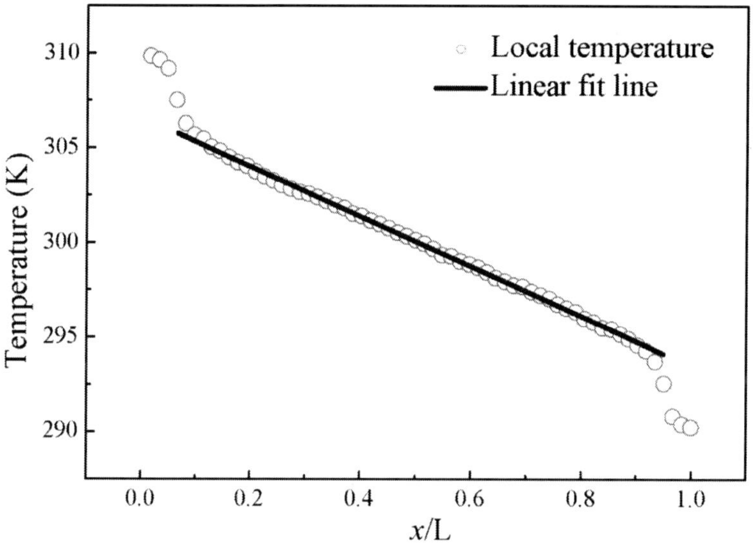

Figure 1.5 Temperature profile of SiNWs in NEMD simulation. The longitudinal direction is set along x-axis. The hollow circle plots the local temperature of each layer along the longitudinal direction, and the solid line is the linear fit line of the local temperature.

NEMD simulation is usually performed long enough to allow the system to reach the non-equilibrium steady state where the temperature gradient is well established (Fig. 1.5) and the heat flux going through the system is time independent. Thermal conductivity is calculated according to Fourier's law of heat conduction

$$\kappa = -\frac{J}{\nabla T}, \tag{1.34}$$

where ∇T is the temperature gradient that can be calculated from the linear fit line of temperature profile (Fig. 1.5), and J is the heat flux transported along the nanowire. In NEMD simulations, the heat flux can be calculated according to the energy injected into/extracted from the heat source/sink across unit area per unit time. In a non-equilibrium steady state, these two rates are equal.

1.5.2 *Effects of Heat Bath*

In both schemes of heat baths introduced in Section 1.4.1, there is a free parameter, namely τ in NH heat bath and λ in Langevin heat bath, which controls the strength of the noise in heat bath. In this section, we take SiNWs as an example to study the impact of heat bath on the calculated thermal properties of homogeneous materials [32]. Moreover, we extend our study to heterogeneous materials, such as Si/Ge NW junctions, in which a rectification of heat current in different directions is of particular interest. Velocity Verlet algorithm is used to numerically integrate Newton's equation of motion. To derive the force term, SW potential is used. The temperature of hot and cold heat baths are set as 310 K and 290 K, respectively. Simulations are performed long enough to allow the system to reach a non-equilibrium steady state where the heat flux going through the system is time independent. All results given in this section are obtained by averaging about 5×10^7 time steps, and each time step is set as 0.8 fs. Free boundary condition is used to atoms on the outer surface of the nanowires, and fixed boundary condition is imposed on the boundary layers at two ends of the nanowires. Thermal conductivity is calculated according to Eq. 1.34.

We first study thermal conductivity of [100] SiNWs with a cross-section of 3×3 unit cells (lattice constant is 0.543 nm, eight atoms in each unit cell) and 10 unit cells in the longitudinal direction. In Fig. 1.6 we show the effect of the number of heat bath layers (NL) on the thermal properties of SiNWs. Here $\tau = 0.1$ and $\lambda = 10$ are used in NH heat bath and Langevin heat bath, respectively. A linear temperature gradient is always observed in the interior, and the main difference in temperature profile between different heat bath types is the temperature jump between the heat bath layers and interior layers. With only one layer of NH heat bath, there exists a large temperature jump (TJ) between the heat bath layer and its neighboring layer (Fig. 1.6a), while the temperature jump is much smaller with one layer of Langevin heat bath (Fig. 1.6b).

This temperature jump can be explained by localized edge mode (LEM) of phonons. With fixed boundary condition, there exists edge mode localized at the neighboring layer next to the fixed boundary [33]. This edge mode is actually a quite generic feature of materials

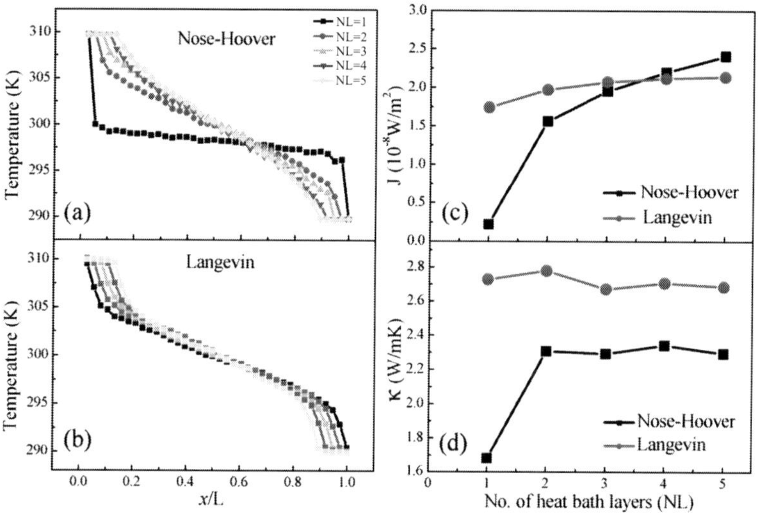

Figure 1.6 Impacts of the number of heat bath layers (NL) on thermal properties of SiNWs. (a) Temperature profile with different NL of Nosé–Hoover heat bath. (b) Temperature profile with different NL of Langevin heat bath. (c) Heat flux J versus NL. (d) Thermal conductivity κ versus NL.

in thermal transport and essentially originated from the specific geometrical configuration of the edge region [34], very similar to the electronic and/or spin edge states [35–37]. It should be emphasized that localization effect is a quite generic consequence of the breaking of spatial periodicity due to the imposed boundary condition in finite system. Therefore, LEM also exists with other boundary conditions in finite system (e.g., free or periodic) [33].

Due to the localization, LEM contributes little to heat transport. If heat bath is applied to this region, LEM will be excited and localized at this region, while other modes can propagate and contribute to the heat transport. We can see from Eqs. 1.25 and 1.26 that in order to maintain a constant temperature, the mechanism of NH heat bath is to introduce a viscosity force, which is proportional to the velocity, and the proportionality is determined from the velocities of all the particles. Due to this deterministic characteristic of NH heat bath, once LEM is excited, it will accumulate at the heat bath layer over time, because LEMs have a larger amplitude than other modes and account for a large percentage of the total modes [33]. As a result, although the heat bath can maintain a constant temperature, the major contribution comes from LEM, which cannot be utilized in heat transport. Therefore, there exists a large TJ when there is only one layer of NH heat bath. With the number of heat bath layer increasing, TJ will be reduced as shown in Fig. 1.6a. This is because of the exponential decay feature of LEM over distance. With multiple layers of heat bath which are away from the localized region, other modes can be excited and dominate in heat transport. Therefore, the localized effect on the thermal transport is suppressed. The reduction in TJ leads to large increase in heat flux from NL = 1 to NL = 2 (Fig. 1.6c). Further increase of heat bath layers cannot eventually eliminate TJ. The small retained TJ is due to the thermal interface resistance between heat bath and rest of the parts.

With Langevin heat bath, the accumulation effect of LEM can be removed automatically, resulting in a small TJ even with one layer of heat bath (Fig. 1.6b). In addition, when NL increases, heat flux converges to a constant value much faster than the case with NH heat bath. The reason is that with Langevin heat bath, each mode can be excited randomly at every time step due to the stochastic characteristic of Langevin heat bath, which can effectively suppress the accumulation of LEM over time.

The large TJ induced by LEM can be further understood by looking at the autocorrelation function of atom's velocity. Figure 1.7 plots the normalized autocorrelation function of velocity for atoms in the middle of NWs with the same parameters used in Fig. 1.6a,b. With one layer of NH heat bath, since it mainly utilizes LEM to maintain the constant temperature, there exists obviously nonvanishing correlation in the long-time region. This artificial correlation is the direct evidence of the existence and accumulation effect of LEM mentioned above, which leads to the large TJ. With multiple layers of NH, the artificial correlation vanishes because all the modes can contribute to the thermal transport, thus a much smaller TJ. Moreover, due to the stochastic characteristic of Langevin heat bath, the artificial correlation does not exist, regardless of the number of heat bath layers applied.

Next we check the temperature jump in other popularly used deterministic heat bath, such as the Berendsen heat bath [38]. It is a

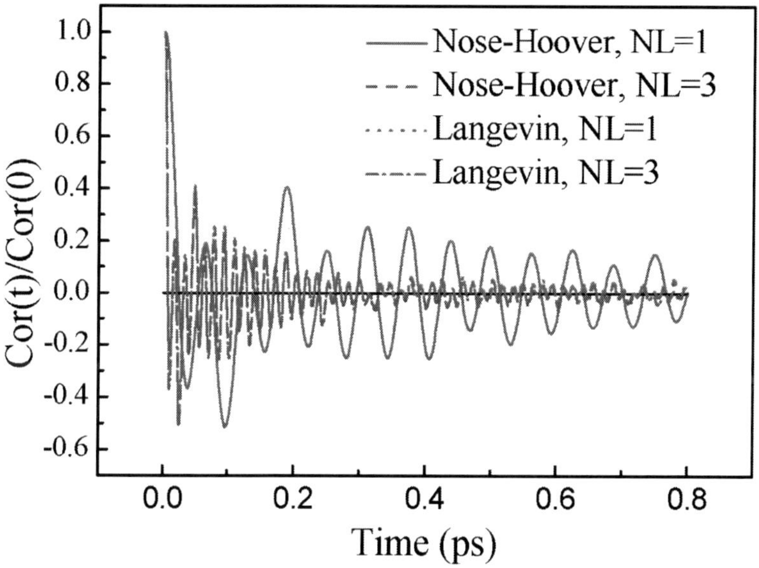

Figure 1.7 Normalized velocity autocorrelation function with different number of layers (NL) of Nosé–Hover and Langevin heat bath applied. Except for the solid line, the other three lines almost overlap with each other.

velocity-scaling type heat bath, with the scaling factor [38]

$$\gamma = \left[1 + \frac{\Delta t}{\tau_{\mathrm{T}}}\left(\frac{T_0}{T} - 1\right)\right]^{1/2}, \tag{1.35}$$

where Δt is the time step, τ_{T} is the relaxation time, T_0 and T are the aimed temperature and instantaneous temperature, respectively. The relaxation time τ_{T} should be properly chosen to avoid unrealistically low-temperature fluctuations with small τ_{T}, and the inactive sampling with large τ_{T} (e.g., $\tau_{\mathrm{T}} \to \infty$) [39]. Based on such considerations, we set $\tau_{\mathrm{T}} = \Delta t \times 10^4$ (8 ps) in our study. As shown in Fig. 1.8, with Berendsen heat bath, the NW fails to reach the aimed temperature (310 K and 290 K) at two ends. More seriously, large TJ persists regardless of the number of heat bath layers applied. This even poorer performance of Berendsen heat bath is a consequence of its inability to reproduce canonical ensemble, because of the artificial velocity-scaling scheme [39], which has been reported to cause artifacts in various studies

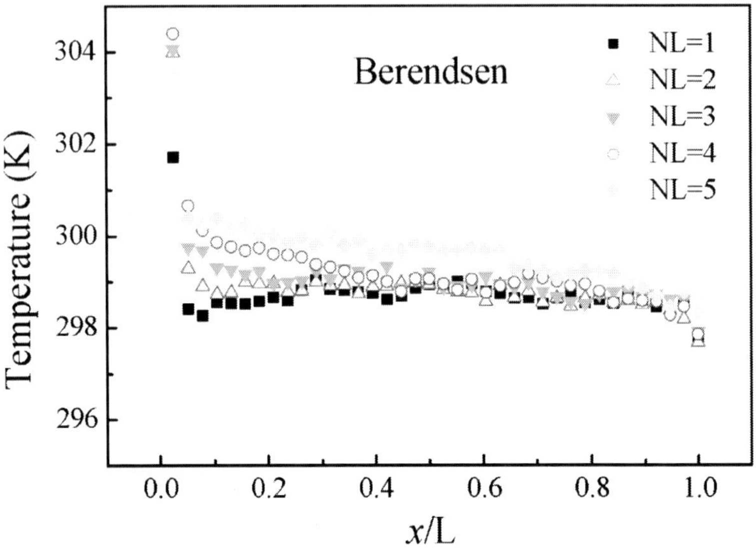

Figure 1.8 Temperature profile with different number of layers (NL) of Berendsen heat bath applied.

[40–43]. Therefore, in the following section, we mainly concentrate on the parameter effect of both NH and Langevin heat bath.

From the calculated heat flux and temperature gradient, we can obtain the thermal conductivity. From Fig. 1.6d, it is obvious that with Langevin heat bath, one can obtain a consistent result regardless of the number of heat bath layers applied. However, due to the existence of LEM and the deterministic nature of NH heat bath, multiple heat bath layers are required in order to get a consistent result. In both NH and Langevin heat bath, the calculated thermal conductivity converges to a constant when NL $\geq$ 3. However, the converged values of thermal conductivity are different with different heat baths. Moreover, there is a free heat bath parameter, τ/λ, which effectively controls the strength of noise (response time in NH heat bath/dissipate rate in Langevin heat bath). Since we fix the heat bath parameter in the first part of this section, whether the choice of free parameter in different heat bath is equivalent to each other and how it can influence the calculated thermal properties are yet not clear. In the following part, we fix the heat bath temperature at 310 K and

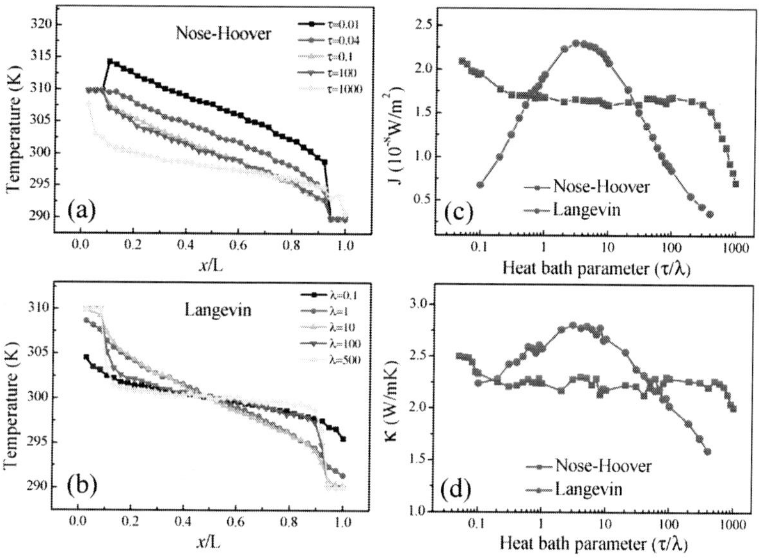

Figure 1.9 Impacts of heat bath parameter on thermal properties of SiNWs. The nanowire has a fixed cross-section of 3×3 unit cells. In the longitudinal direction, it has a fixed length of 10 unit cells with three layers of heat bath at each end. (a) Temperature profile with parameter τ of Nosé–Hoover heat bath. (b) Temperature profile with parameter λ of Langevin heat bath. (c) Heat flux J versus heat bath parameter. (d) Thermal conductivity κ versus heat bath parameter. Here $0.05 \leq \tau \leq 1000$, and $0.1 \leq \lambda \leq 400$.

290 K, and tune the heat bath parameters to study their impacts on heat flux, temperature profile, and thermal conductivity.

Next we set NL $=$ 3 for both NH and Langevin heat bath. In Fig. 1.9 we plot the impacts of heat bath parameter on the calculated thermal properties of SiNWs. For NH heat bath, τ cannot be too small (e.g., $\tau = 0.01$ in Fig. 1.9a) since it produces wrong temperature profile in this case, although the heat bath still can reach the aimed temperature and a temperature gradient can be established in the middle region. With the increase of τ from 0.01, the temperature gradient in the middle region is almost the same, but the mean temperature decreases until $\tau = 0.2$. This results in a decrease in heat current when $\tau < 0.2$. Particularly when $\tau = 0.04$, TJ at the left boundary is nearly zero, while TJ at the right boundary is about 5 K. This should be a numerical artifact because TJ has its physical origin from the thermal interface resistance between heat bath and the middle region, thus cannot be eliminated by tuning parameter. With further increase of $\tau(0.2 < \tau < 300)$, temperature profile becomes correct and stable: both heat flux and temperature gradient become insensitive to τ, thus the calculated thermal conductivity in Fig. 1.9d is almost a constant with small fluctuation. Next, when $\tau > 300$, temperature gradient becomes smaller and heat flux decreases quickly. This is because in the large τ range, the distribution of ς becomes a δ-function [27]. As a result, it would require longer simulation times in order to ensure the decay of correlations. This prevents the heat bath from reaching the aimed temperature due to the limited simulation time, and also a smaller temperature gradient. Therefore, from the consideration of a practical computational time, one should not choose an extremely large τ.

For Langevin heat bath, in the weak damping limit $\lambda \to 0$ (e.g., $\lambda = 0.1$ in Fig. 1.9b), the heat bath cannot reach the aimed temperature. However, in the strong damping limit (e.g., $\lambda = 500$ in Fig. 1.9b), although the heat bath can reach the aimed temperature, large TJ is observed at the boundary. In both cases, small temperature gradient is generated, which induces a small heat current. In the middle range of λ, a temperature profile with correct heat bath temperature and small TJ can be established. This causes heat flux and thermal conductivity first increase then decrease with the increase of λ as shown in Fig. 1.9c, d.

In MD calculation, it must be confirmed that the temperature profile is correct and a temperature gradient can be well established in order to achieve accurate prediction results. Based on these considerations and the results shown in Fig. 1.9, an intermediate value of λ (from 1 to 100) is recommended for Langevin heat bath. For NH heat bath, Fig. 1.9 suggests τ from 0.2 to 300 is the optimal choice in numerical simulation when NH heat bath is applied. Moreover, we can see from Fig. 1.9d that the discrepancy of thermal conductivity in Fig. 1.6d is mainly due to the choice of heat bath parameter: the same result can be obtained if the parameter is chosen properly.

In addition to homogeneous materials, heterogeneous materials such as carbon nanotube based [44] and graphene based [45] nano-junctions are promising candidates for thermal rectifier application.

In these systems, it is problematic to define thermal conductivity according to Fourier's law, due to large TJ at the interface of two different materials. Instead, because of the asymmetry of the heterogeneous materials, people are more interested in the rectification effect of heat flow in such materials [46], namely the difference between the heat flux in different directions. In the following section, we extend our study to heterogeneous materials and discuss the effect of heat bath parameter on the thermal rectification.

We use Si/Ge nanojunction as an example. It has a fixed cross-section of 3×3 unit cells, 5 unit cells of Si and 5 unit cells of Ge in the longitudinal direction. Three layers of heat bath are applied at each end, with heat bath temperature set at 310 K and 290 K. Here we define J_+ (J_-) to be the heat flux of non-equilibrium steady state when Si (Ge) end is attached to the high-temperature heat bath. We define the rectification efficiency *RE* to be

$$RE = \frac{J_+ - J_-}{J_-}. \tag{1.36}$$

Figure 1.10a,b shows the dependence of heat current on heat bath parameters. It has been checked that temperature profiles with these parameters are similar to that shown in Fig. 1.10c for both NH and Langevin heat bath. With NH heat bath, J_+ and J_- have the distinct dependence on τ. In the small τ limit, J_+ is much larger than J_-. With the increase of τ, J_+ drops rapidly and finally converges to a small value, while J_- first increases, then decreases, and finally converges to a value that is slightly larger than J_+. As a result, there exists a large value of *RE* in the small τ limit as shown in Fig. 1.10d. Moreover, *RE* changes from positive to negative when $0.05 \leq \tau \leq 100$. Thus NH heat bath fails to give a consistent result. In contrast, with Langevin heat bath, Fig. 1.10b shows the same dependence on λ for both J_+ and J_-, and there is only a small difference between them. This results in a small value of *RE* as shown in Fig. 1.10d for Langevin heat bath. More importantly, Langevin heat bath can produce a consistent result ($RE < 0$) that J_- is always larger than J_+, regardless of the heat bath parameter λ. Si–Ge nanowire is a mass graded nanojunction. The results calculated with Langevin heat bath and NH heat bath with large $\tau(\tau > 1)$ suggest that the heat flux runs preferentially along the direction of decreasing mass.

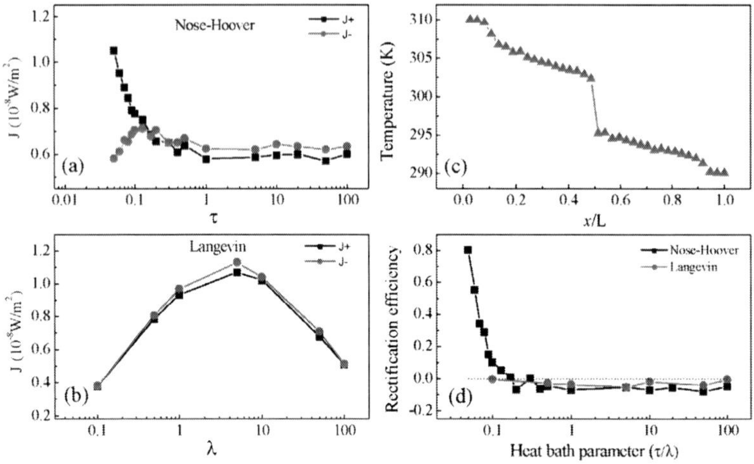

Figure 1.10 Thermal rectification effect in Si/Ge nanojunction versus heat bath parameters. The square and circle denote the results for Nosé–Hoover (τ) and Langevin (λ) heat bath, respectively. (a) $J_\pm$ versus parameter τ of Nosé–Hoover heat bath. (b) $J_\pm$ versus parameter λ of Langevin heat bath. (c) Typical temperature profile in Si/Ge nanojunction. (d) Rectification efficiency versus heat bath parameter.

This conclusion is consistent with experimental result by Chang *et al.* [47]. They demonstrated the thermal rectification effect in carbon and boron nitride nanotubes that were inhomogeneously mass-loaded with heavy molecules. A larger heat flow was observed when heavy-mass end is at higher temperature.

Here we have discussed the impacts of heat bath on calculated thermal properties of nanostructures in non-equilibrium molecular dynamics simulations. Due to the existence of localized edge modes and their accumulation effect induced by the deterministic characteristic of Nosé–Hoover heat bath, multiple layers of Nosé–Hoover heat bath are required in order to reduce the temperature jump at the boundary. Even with one layer of Langevin heat bath, it can eliminate the accumulation of localized edge modes due to its stochastic excitation of all modes, giving rise to a small temperature jump at the boundary. In addition, in order to obtain the correct temperature profile, intermediate values of heat bath parameter,

$0.2 < \tau < 300$ for Nosé–Hoover heat bath and $1 < \lambda < 100$ for Langevin heat bath, are recommended.

1.5.3 *Some Applications*

The thermal conduction is described by Fourier's law, where thermal conductivity is defined as one intrinsic property of a material. For a certain material with definite composition and structure, the thermal conductivity is believed to be an intensive property that should be independent on the size and geometry. This belief has received great success in describing macroscopic heat transport in the past two hundred years.

However, a rigorous proof for Fourier's law from microscopic Hamiltonian dynamics is still lacking, even though it is already 200-year old. Therefore, it is still an open and much debated question whether Fourier's law is valid or applicable in low-dimensional systems. In past two decades, size dependent thermal conductivity has already been observed in many theoretical models. The discovery of this anomalous behavior in general low-dimensional model has inspired enormous studies. Low-dimensional nanoscale systems have been extensively studied due to their promising potential applications for future electronic, optoelectronic, and phononic/thermal devices. Thus these low-dimensional nanoscale systems are ideal platforms to verify the fundamental physics.

By using non-equilibrium molecular dynamics simulation, Zhang and Li [48] studied thermal conductivity of single-walled carbon nanotubes (SWNTs). It is clear that, unlike its electronic counterpart, the thermal conductivity of SWNTs does not depend on the chirality both at low temperature and room temperature. The thermal conductivity κ versus the tube length L is shown in log–log scale in Fig. 1.11 for both (5,5) and (10,10) SWNTs at 300 K and 800 K, respectively. It is interesting to see that thermal conductivity κ diverges with SWNT length L as $\kappa \sim L^{\beta}$, and the value β depends on temperature as well as tube radius. For a SWNT, β decreases as temperature increases; and at the same temperature, β decreases as the tube radius increases. This can be qualitatively explained by the mode coupling theory. At high temperature, the transverse vibrations are much larger than that at low temperature, thus the

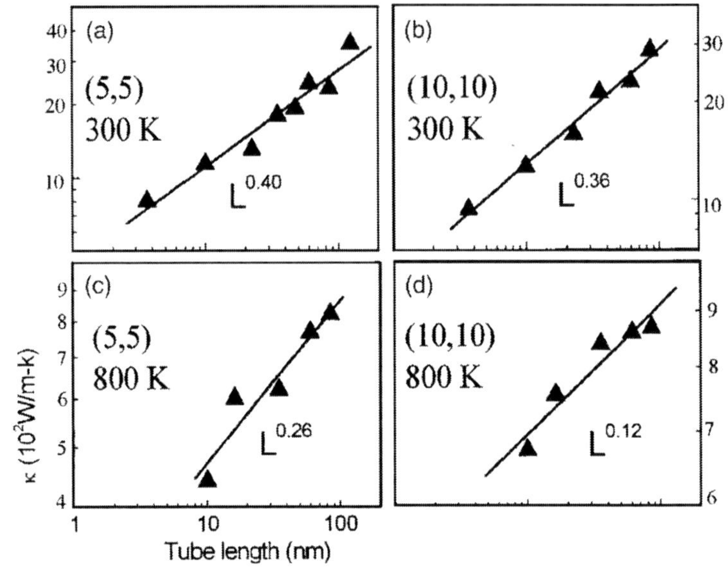

Figure 1.11 The thermal conductivity κ versus the tube length L in log–log scale for both (5,5) and (10,10) SWNTs at 300 K and 800 K, respectively.

interaction between the transverse modes and longitudinal modes becomes stronger, which leads to a smaller value of β. For the (5,5) tube at $T = 300$ K, because the small tube radius, the coupling between transverse and longitudinal mode is still very weak, this leads to the thermal conduction behavior is very close to that one in the 1D Fermi–Pasta–Ulam-type lattice, namely, the thermal conductivity κ diverges with L, as $L^{0.4}$.

The length dependent thermal conductivity was predicted by MD simulation, then it was verified experimentally [49]. This shows that MD simulation does not rely on any thermodynamic limit assumptions and it thus is applicable for any system size, which is important for the study of thermal conductivity of realistic nanoscale systems.

The high thermal conductivity of SWNTs is for a pure and defect free tube. Actually, nanotubes can have natural defects and doping in the process of fabrication. Doping with isotopes and other atoms has played key roles in some of the most important problems of the materials. Since the phonon frequency depends on mass,

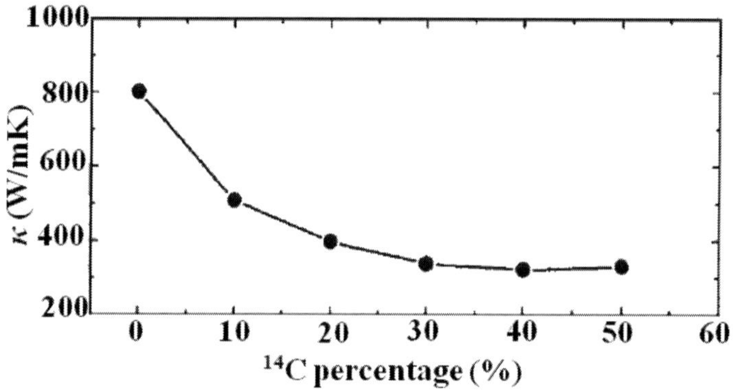

Figure 1.12 Thermal conductivity versus ^{14}C impurity percentage for a (5,5) SWNT at 300 K.

isotopic doping can lead to increased phonon scattering. From NEMD simulation, the isotopic effect on thermal conductivity of SWNT was calculated [48], as shown in Fig. 1.12. It is found that the thermal conductivity decreases as the percentage of ^{14}C impurity increases. With 20% ^{14}C, the thermal conductivity is reduced to about 50% of that one in a pure ^{12}C SWNT. The thermal conductivity decreases more quickly at low percentage range than at higher range. From this curve, thermal conductivity decreases about 20% with only 5% ^{14}C isotope impurity. This result tells us that one can modulate the thermal conductivity of carbon nanotubes by adding isotope impurity as it alters only thermal conductivity and has no effect on the electronic properties. This MD predicted isotopic effect on thermal conductivity has been verified experimentally by Chang *et al.* [50]. In their experiment, they found that 20% isotopic doping can induce 50% reduction in thermal conductivity of nanotube. This is with quantitative agreement with the MD calculations.

Very recently, interesting thermal conductivity dependence in Si-coated Ge nanowire (NW) was found by using NEMD simulation [51]. The structure of the Ge/Si core–shell NW is shown in Fig. 1.13a. Here the cross-sectional side length (diameter) of Ge NW D_{Ge} changes from 2.8 to 11.3 nm. Figure 1.13b shows the room temperature thermal conductivity of GeNWs ($D_{\mathrm{coating}} = 0$) and Ge/Si core-shell NWs ($D_{\mathrm{coating}} > 0$) for different D_{Ge}. It is obvious

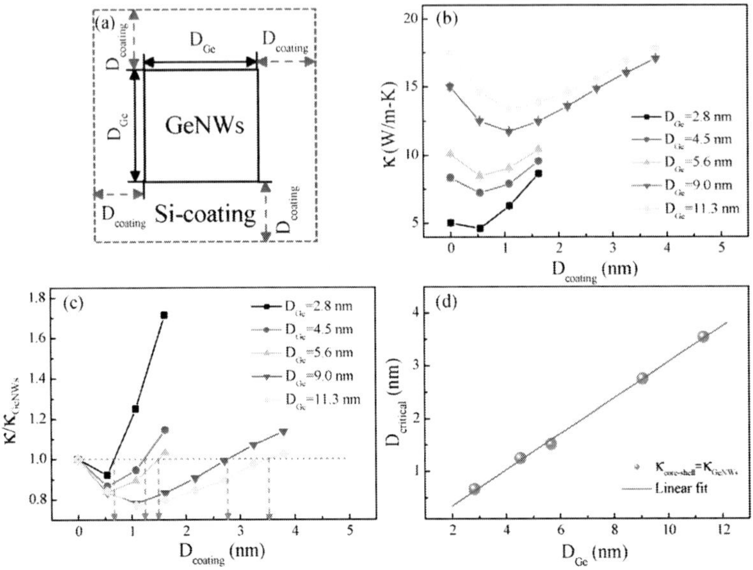

Figure 1.13 Coating configuration and thermal conductivity. (a) Cross-section of GeNWs with Si-coating shell (dashed line). (b) Thermal conductivity of GeNWs ($D_{\text{coating}} = 0$) and Ge/Si core-shell NWs ($D_{\text{coating}} > 0$) for different D_{Ge} at room temperature. (c) Normalized thermal conductivity versus coating thickness for different D_{Ge}. The dashed arrows point the critical coating thickness when thermal conductivity of Ge/Si core-shell NWs ($\kappa_{\text{core-shell}}$) is equal to that of GeNWs (κ_{GeNWs}). The dashed line is drawn to guide the eye. (d) The critical coating thickness D_{critical} versus cross-sectional side length of GeNWs. The solid line draws the linear fit line.

that for each GeNW, the coating of Si atomistic layers can reduce the thermal conductivity. Further coating results in an increase of thermal conductivity.

Increasing coating thickness has two opposite effects on thermal conductivity. On the one hand, the creation of core-shell structures will induce the phonon resonance between the transverse and longitudinal modes, thus offering a coherent mechanism to reduce thermal conductivity. In Si-coated GeNW, atoms on the same cross-section plane have different sound velocity in the longitudinal direction. As a result, atoms near the interface are stretched, which induces a strong coupling between the transverse and longitudinal motions. Because of the spatial confinement on the cross-section plane, the transverse phonons are quantized and nonpropagating. The coupling between the transverse and longitudinal motions results in the resonance when the frequency of the longitudinal phonon mode is close to the eigenfrequency of the transverse mode. This resonance effect results in longitudinal phonon localization, thus offering a coherent mechanism to reduce thermal conductivity.

On the other hand, the coating layers increase the entire cross-section area. Because of the increase of the entire cross-section area after coating, the strength of the surface scattering is reduced, leading to the increase of thermal conductivity with the increase of the coating thickness. Therefore, thermal conductivity of Si-coated GeNWs is determined by these two competing effects. When the coating thickness is less than a certain critical value, the suppression of the longitudinal phonon transport is the dominating factor, corresponding to the reduction of thermal conductivity. However, when the coating thickness is greater than a certain critical value, the reduced surface scattering dominates the thermal transport, leading to the fact that thermal conductivity of the resultant Ge/Si core-shell NWs becomes larger than that of pristine GeNW without coating.

These results show that in practical application one can control thermal conductivity of NWs by coating other atoms. This approach offers novel avenues and more flexibility for the design and thermal management in nanostructures. For instance, small diameter NWs are favorable for thermoelectric applications due to the low thermal conductivity but are more challenging for experimental synthesis. Through the coating method, the low thermal conductivity feature close to the very thin NW can be obtained from a much thicker NW. This point can be seen from Fig. 1.13b. Thermal conductivity of pristine GeNWs with $D_{\mathrm{Ge}} = 9.0$ nm is close to that of coated GeNWs with $D_{\mathrm{Ge}} = 11.3$ nm and $D_{\mathrm{coating}} = 2.7$ nm. The resultant core-shell NW has a cross-sectional side length of 16.7 nm, which is almost twice of the cross-sectional side length for pristine GeNW with the similar thermal conductivity.

Moreover, for NWs with a given diameter, the coating method can preserve the low thermal conductivity feature of the original NWs up to certain critical coating thickness. For example, for NW with $D_{\mathrm{Ge}} = 9.0$ nm, coating layers less than 2.7 nm can give rise to thermal conductivity lower than that of pristine GeNW. As this coating thickness is achievable with atomistic layer deposition technology, this MD-based study indicates a practical approach that can be realized to tune thermal conductivity.

It is worth to point out the critical coating thickness depends on the diameter of the NW before coating. To see this dependence more clearly, the normalized thermal conductivity versus coating thickness for GeNWs with different diameters was plotted in Fig. 1.13c. Thermal conductivity of GeNW is used as reference at each D_{Ge}. According to the abovementioned two competing effects, the maximum reduction of thermal conductivity occurs at the optimal coating thickness of about 0.5–1 nm. Further increase of coating thickness will result in the increase of thermal conductivity, but the increase trend becomes slower at larger diameters. The dashed line in Fig. 1.13c draws the unity for reference, and the arrows point the critical coating thickness below which coating is effective for the purpose of thermal conductivity reduction.

Figure 1.13d shows the dependence of critical coating thickness D_{critical} on the NW diameter: it monotonically increases with NW diameter, with a linear slope of about 0.34. These MD simulation results suggest that coating is a quite effective method for thermoelectric applications and is more robust at larger diameter.

1.6 Equilibrium Molecular Dynamics

1.6.1 *Green–Kubo Formula*

Equilibrium molecular dynamics (EMD) simulation is based on the Green–Kubo formula (GKF) derived from the fluctuation–dissipation theorem [28] and linear response theory, which relates thermal conductivity with heat current autocorrelation function (HCACF) [52–54]. In GKF, the entire dynamics of the systems is expressed through the time correlation function in thermal equilibrium, which invokes little assumption about the physical property of the material. EMD simulation has been applied to study thermal properties of various materials [55–58].

Several derivations of GKF exist in literature [59]. Here we follow the derivation by Kubo *et al.* [60], with emphasis on some essential steps. There is also a Kubo formula for electrons [53], which relates electrical conductivity to the electrical current autocorrelation function. For the electrical case, there is a mechanical disturbance term [60] in the Hamiltonian, which describes the energy perturbation due to the applied external electrical field. The derivation of electrical Kubo formula follows the standard perturbation theory. For the thermal case, however, the derivation of GKF is less straightforward but relies on the statistical hypothesis of local thermal equilibrium, which can be described by the local space-dependent temperature $T(\mathbf{r}) = [k_B\beta(\mathbf{r})]^{-1}$.

Consider a system with Hamiltonian H and volume V at temperature $T(\mathbf{r})$ subjected to a small thermal disturbance $\delta T(\mathbf{r})$. The local thermal equilibrium holds and can be described by the local equilibrium density matrix:

$$\rho = \exp\left[-\int d^3\mathbf{r}\beta(\mathbf{r})h(\mathbf{r})\right]/Z, \tag{1.37}$$

where Z is the partition function, and $h(\mathbf{r})$ is the Hamiltonian density operator related to the Hamiltonian as $H = \int d^3\mathbf{r}h(\mathbf{r})$. Similarly, the heat current density operator $\mathbf{s}(\mathbf{r})$ is defined as

$$\mathbf{S} = \int d^3\mathbf{r}\mathbf{s}(\mathbf{r}), \tag{1.38}$$

where $\mathbf{S}$ is the total heat current operator. Due to the energy conservation, these two density operators are related through

$$\frac{\partial h(\mathbf{r})}{\partial t} + \nabla \cdot \mathbf{s}(\mathbf{r}) = 0. \tag{1.39}$$

Without $\delta T(\mathbf{r})$, the system is in thermal equilibrium so that $T(\mathbf{r}) = T_0 = [k_B\beta_0]^{-1}$ is a constant, and there is no net heat current ($\mathbf{J} = \mathrm{Tr}[\rho_0\mathbf{S}] = 0$). After applying the small thermal disturbance, the density matrix becomes

$$\rho = \exp\left[-\beta(H + H')\right]/Z, \tag{1.40}$$

where H' is the small perturbation to the Hamiltonian caused by the thermal disturbance. There is a fundamental equality in quantum mechanics for any two operators $\hat{a}$ and $\hat{b}$ [60] so that

$$e^{\beta(\hat{a}+\hat{b})} = e^{\beta\hat{a}}\left(1 + \int_0^\beta d\lambda e^{-\lambda\hat{a}}\hat{b}e^{\lambda(\hat{a}+\hat{b})}\right), \tag{1.41}$$

where λ is a phase space variable. With this equality, the numerator of density matrix in Eq. 1.40 can be approximated to the first order as

$$e^{-\beta(H+H')} = e^{-\beta H}\left(1 - \int_0^{\beta} d\lambda e^{\lambda H} H' e^{-\lambda H}\right). \tag{1.42}$$

The small perturbation to the Hamiltonian can be expressed as

$$\begin{aligned} H' &= \frac{1}{T}\int d^3\mathbf{r}\delta T(\mathbf{r})h(\mathbf{r}) \\ &= -\frac{1}{T}\int dt \int d^3\mathbf{r}\delta T(\mathbf{r})\nabla\cdot\mathbf{s}(\mathbf{r}) \\ &= \frac{1}{T}\int dt \int d^3\mathbf{r}\nabla T(\mathbf{r})\cdot\mathbf{s}(\mathbf{r}) \\ &= \frac{1}{T}\int dt\nabla T(\mathbf{r})\cdot\mathbf{S}, \end{aligned} \tag{1.43}$$

where we have used the integrated form of Eq. 1.39, performed an integration by part, and assumed the temperature gradient to be a spatial constant.

With the small thermal disturbance, the net heat current becomes

$$\mathbf{J} = \mathrm{Tr}[\rho\mathbf{S}] = -\frac{1}{TV}\int_0^{\infty} dt \int_0^{\beta} d\lambda \left\langle e^{\lambda H}\left(\nabla T\cdot\mathbf{S}(0)\right)e^{-\lambda H}\mathbf{S}(t)\right\rangle, \tag{1.44}$$

where the angular bracket denotes the ensemble average. According to the Fourier's law

$$\kappa_{\mu\upsilon} = -\frac{J_\mu}{\nabla_\upsilon T}, \tag{1.45}$$

where μ and υ are two Cartesian indices, thermal conductivity can be written in the tensor form as

$$\begin{aligned} \kappa_{\mu\upsilon} &= \lim_{\tau\to\infty}\lim_{V\to\infty}\frac{1}{TV}\int_0^{\tau} dt \int_0^{\beta} d\lambda \left\langle e^{\lambda H} S_\upsilon(t') e^{-\lambda H} S_\mu(t'')\right\rangle \\ &= \lim_{\tau\to\infty}\lim_{V\to\infty}\frac{1}{TV}\int_0^{\tau} dt \int_0^{\beta} d\lambda \left\langle e^{\lambda H} S_\upsilon(0) e^{-\lambda H} S_\mu(t)\right\rangle \\ &= \lim_{\tau\to\infty}\lim_{V\to\infty}\frac{1}{TV}\int_0^{\tau} dt \int_0^{\beta} d\lambda \left\langle S_\upsilon(-i\hbar\lambda) S_\mu(t)\right\rangle, \end{aligned} \tag{1.46}$$

where we have assumed that in the equilibrium steady state HCACF in Eq. 1.46 only depends on the time difference ($t = t'' - t'$), and used the time dependence of operator in Heisenberg picture: $\hat{a}(t) = e^{i\frac{H}{\hbar}t}\hat{a}(0)e^{-i\frac{H}{\hbar}t}$. Notice that the order of the limits in Eq. 1.46 is important: the volume limit $V \to \infty$ must be taken first, followed by the time limit $\tau \to \infty$ [61]. For the sake of simplicity, we will omit the limits in Eq. 1.46 and put the time limit to infinity in the following discussions.

In the classical limit $\hbar \to 0$, Eq. 1.46 can be further simplified to

$$\begin{aligned} \kappa_{\mu\upsilon} &= \frac{1}{TV}\int_0^{\infty} dt \int_0^{\beta} d\lambda \left\langle S_\upsilon(0)S_\mu(t)\right\rangle \\ &= \frac{1}{k_B T^2 V}\int_0^{\infty} dt \left\langle S_\upsilon(0)S_\mu(t)\right\rangle. \end{aligned} \tag{1.47}$$

For bulk materials with cubic symmetry, thermal conductivity is usually expressed in scalar form as

$$\kappa = \frac{1}{3k_B T^2 V}\int_0^{\infty} dt \left\langle \mathbf{S}(0)\cdot\mathbf{S}(t)\right\rangle, \tag{1.48}$$

in which thermal conductivity is averaged over three diagonal terms. For low-dimensional materials without cubic symmetry, such as nanowire and nanotube, thermal conductivity should be calculated according to Eq. 1.47.

According to Hardy's formulation [62], heat current can be defined as [63]

$$\mathbf{S} = \frac{d}{dt}\sum_i \mathbf{r}_i(t)\varepsilon_i(t), \tag{1.49}$$

where $\mathbf{r}_i(t)$ and $\varepsilon_i(t)$ denote the time-dependent coordinate and total energy of atom i, respectively. Take SW potential as an example, heat current can be further written as

$$\mathbf{S} = \sum_i \mathbf{v}_i\varepsilon_i + \frac{1}{2}\sum_{\substack{i,j\\ i\neq j}} \mathbf{r}_{ij}(\mathbf{F}_{ij}\cdot\mathbf{v}_i) + \frac{1}{6}\sum_{\substack{i,j,k\\ i\neq j,\, j\neq k}} (\mathbf{r}_{ij}+\mathbf{r}_{ik})(\mathbf{F}_{ijk}\cdot\mathbf{v}_i), \tag{1.50}$$

where $\mathbf{v}_i$ is the velocity of atom i, and $\mathbf{F}_{ij}$ and $\mathbf{F}_{ijk}$ denote the two-body and three-body force, respectively. The first term on the right-hand side of Eq. 1.50 describes the heat convection typically occurring in fluids, while the rest terms describe the heat conduction between the atoms, which is dominant part in solids. Therefore, heat current can be simplified as [58]

$$\mathbf{S} = \frac{1}{2}\sum_{\substack{i,j\\ i\neq j}} \mathbf{r}_{ij}(\mathbf{F}_{ij}\cdot\mathbf{v}_i) + \frac{1}{6}\sum_{\substack{i,j,k\\ i\neq j, j\neq k}} (\mathbf{r}_{ij}+\mathbf{r}_{ik})(\mathbf{F}_{ijk}\cdot\mathbf{v}_i), \qquad (1.51)$$

which is computationally more efficient as the calculation of local energy is not required.

1.6.2 *Different Implementations*

The main difficulty in implementing GKF in practical calculations arises from how to carry out the time integral up to infinity. In this section, we review the existing different implementations of GKF in literature. Most of them can be categorized into two types: one is the time-domain approach, and the other is the frequency-domain approach.

The time-domain approach is to handle GKF in time-domain. The most straightforward way is the direct integration method [63], which replaces the integral with summation and numerically records HCACF in time-domain as

$$\begin{aligned} Cor(t) &= \left\langle S_\upsilon(0)S_\mu(t)\right\rangle \\ &= \frac{1}{N-m}\sum_{n=0}^{N-m}\left[S_\upsilon(nt_0)S_\mu((n+m)t_0)\right], \end{aligned} \qquad (1.52)$$

where t_0 is the time step, N is the total number of time steps in EMD simulations, and $m = t/t_0$ is the integer number for time t. In this method, the infinite integral in GKF is replaced by a summation up to a finite cut-off time τ_c

$$\kappa_{\mu\upsilon} = \frac{t_0}{k_B T^2 V}\sum_{t=0}^{\tau_c} Cor(t), \qquad (1.53)$$

In addition to direct integration, several alternatives have also been proposed by making use of certain statistical properties of HCACF. For instance, Li *et al.* suggested to fit HCACF according to the single exponential function in the time interval $[\tau_1, \tau_2]$ [56]

$$Cor(t) = ge^{-t/\tau_0}, \qquad (1.54)$$

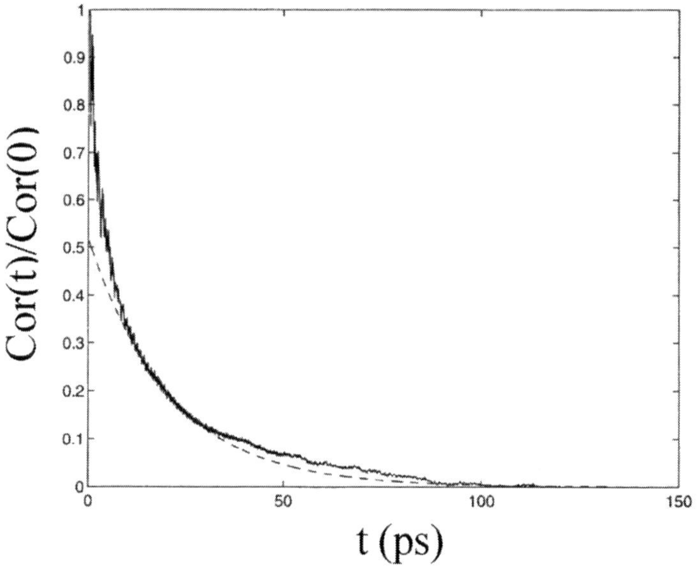

Figure 1.14 Normalized HCACF for perfect β-SiC at 760 K. The dashed line draws the single exponential fit in the range from 10 ps to 30 ps.

where g and τ_0 are two fitting parameters, and thermal conductivity is obtained by analytically calculating the infinite integral in GKF

$$\kappa_{\mu\upsilon} = \frac{g\tau_0}{k_B T^2 V}. \tag{1.55}$$

Figure 1.14 shows an example of the single exponential fit to HCACF for β-SiC adapted from Ref. [56]. In the single exponential fit method, the fitting parameters depend critically on the choice of τ_1 and τ_2, which are selected empirically. This may bring certain degree of ambiguity into the calculation results. Later, Che *et al.* [57] found that due to the different relaxation time for acoustic and optical phonons in diamond, HCACF generally exhibits a two-stage decaying behavior, with a very fast decay at the beginning and followed by a much slower decay in the long-time tail, as shown in Fig. 1.15.

To take into account this two-stage decaying characteristic of HCACF, Che *et al.* [57] proposed to fit HCACF based on the double exponential function, and then calculate thermal conductivity based on the fitting parameters as

$$Cor(t) = A_o e^{-t/\tau_o} + A_a e^{-t/\tau_a}, \tag{1.56}$$

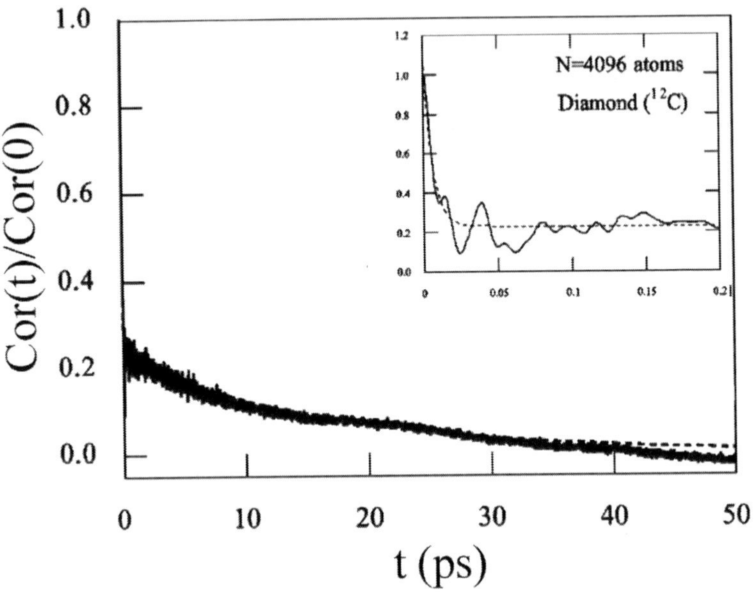

Figure 1.15 Normalized HCACF for the $N = 4096$ atoms simulation of ^{12}C diamond at 300 K. The dashed line draws the double exponential fit, and the inset zooms in for the short-time region.

$$\kappa_{\mu\upsilon} = \frac{A_o \tau_o + A_a \tau_a}{k_B T^2 V}. \tag{1.57}$$

This method removes the ambiguity on the selection of the fitting time interval. Although double exponential function is used for the fitting, Che *et al.* found the high-frequency optical modes contribute only 0.1% to the thermal conductivity of diamond [57].

About the double exponential fit approach, McGaughey *et al.* [64] found that the two-stage decaying characteristic also exists in the mono-atomic systems, such as Lennard–Jones argon. In a more general sense, they found that fast and slow decays are associated with the short-wavelength and long-wavelength phonons, respectively. Moreover, Chen *et al.* [65] found that due to the finite time cut-off, the fitted double exponential function is not fully relaxed to zero but to a tiny nonzero value Y_0, although HCACF itself can decay to zero (Fig. 1.16). The Y_0 term results from

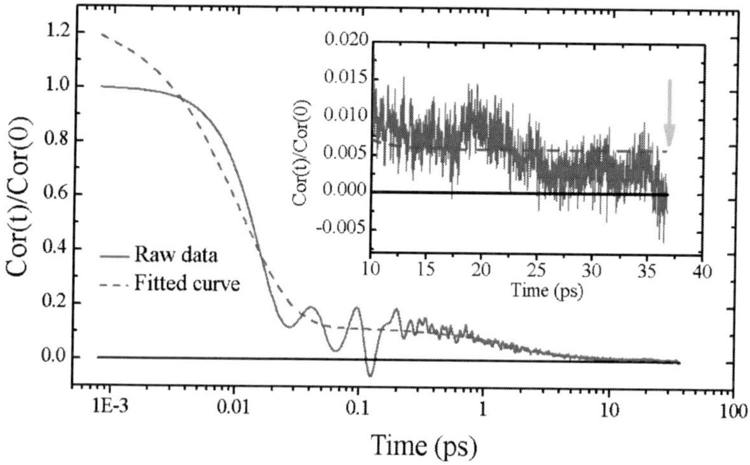

Figure 1.16 Normalized HCACF for bulk Si at 1000 K. The solid and dashed line draws the raw data and double exponential fit, respectively. The arrow denotes the cut-off time.

the contribution of long-wavelength phonons which have a longer relaxation time than the cut-off time.

Chen *et al.* [65] proposed a nonzero correction to the double exponential fit method,

$$Cor(t)/Cor(0) = A_1 e^{-t/\tau_1} + A_2 e^{-t/\tau_2} + Y_0, \qquad (1.58)$$

and correspondingly thermal conductivity is calculated as

$$\kappa = \frac{Cor(0)}{3k_B T^2 V} \left(A_1\tau_1 + A_2\tau_2 + Y_0\tau_c\right), \qquad (1.59)$$

where τ_c is the cut-off time.

Using bulk silicon as an example, Chen *et al.* [65] calculated the thermal conductivity with and without this nonzero correction Y_0. At 1000 K, thermal conductivity of bulk silicon is about 32W/(m-K) [66]. As shown in Fig. 1.17, the saturated value from EMD simulation without Y_0 is only about 16 W/(m-K), much lower than the experimental benchmark. With the improved correction, the MD predicted thermal conductivity agrees well with the experimental value, demonstrates the accuracy of this improved double exponential fitting approach.

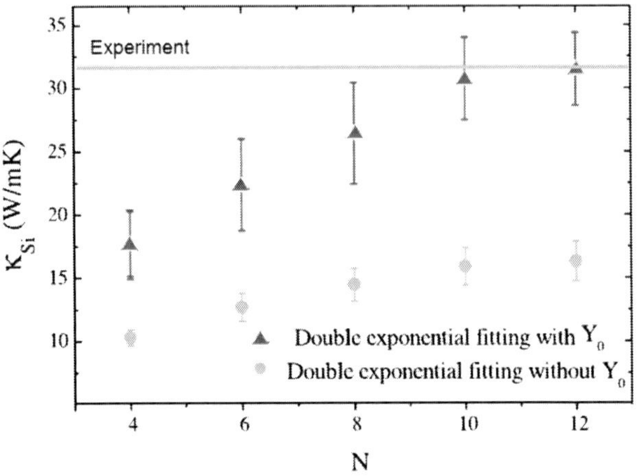

Figure 1.17 Thermal conductivity of the crystalline Silicon at 1000 K versus super cell size ($N \times N \times N$ unit cells) for different methods. The solid line draws the experimental result at the same temperature from Ref. [66].

The frequency-domain approach is an indirect way to handle GKF in frequency-domain by using Fourier transform and Wiener–Khinchin theorem. From mathematical point of view, one can consider GKF in Eq. 1.48 as the Fourier transform of HCACF, and thus can further define spectral thermal conductivity as

$$\kappa(\omega) = \frac{1}{3k_B T^2 V} \int_0^\infty dt \, \langle \mathbf{S}(0) \cdot \mathbf{S}(t) \rangle \, e^{i\omega t}, \tag{1.60}$$

in which the conventional thermal conductivity can be regarded as the spectral thermal conductivity in the static limit $\omega \to 0$. Similarly, spectral heat current can be defined as

$$\mathbf{S}(\omega) = \int_0^\infty dt \mathbf{S}(t) e^{i\omega t}. \tag{1.61}$$

Wiener–Khinchin theorem [67] asserts that in a stationary random process, the power spectral density is the Fourier transform of the corresponding autocorrelation function. Thus, according to Wiener–Khinchin theorem, the heat current power spectral density can be written as

$$|\mathbf{S}(\omega)|^2 = \int_0^\infty dt \, \langle \mathbf{S}(0) \cdot \mathbf{S}(t) \rangle \, e^{i\omega t}. \tag{1.62}$$

Therefore, thermal conductivity can be calculated from the spectral density of heat current as [55]

$$\kappa = \kappa(\omega)|_{\omega=0}$$
$$= \frac{1}{3k_B T^2 V} |\mathbf{S}(\omega)|^2\Big|_{\omega=0}. \tag{1.63}$$

Strictly speaking, thermal conductivity can only be obtained from this method in the static limit, $\omega \to 0$, which is in practice infeasible due to the finite time simulation. Therefore, the static limit is simply approximated by extrapolating the high-frequency data to zero frequency [55]. Later, this spectral method was improved by Volz *et al.* [58] with the assumption of single exponential decay of HCACF in Eq. 1.54. Based on this assumption, the spectral thermal conductivity is then fitted at high frequency according to [58]

$$\kappa(\omega) = \frac{\kappa(0)}{1 + i\omega\tau_0}, \tag{1.64}$$

where $\kappa(0)$ and τ_0 are two fitting parameters, corresponding to the static thermal conductivity and single exponential decay constant, respectively. In this method, the static thermal conductivity is obtained as the fitting parameter, thus no extrapolation is involved.

1.6.3 *Determination of Cut-Off Time*

In EMD simulation, the accuracy of HCACF is limited by the total simulation time, which corresponds to the maximum ensemble average of the correlation function at $t = 0$ (Eq. 1.52). HCACF becomes less accurate over time because of the smaller ensemble average one can get from a finite time simulation. Therefore, numerical error (noise) is inevitably introduced into the calculation, and eventually will contaminate HCACF when it decays to a small value. Consequently, HCACF is only reliable up to a finite time (cut-off time). Thus, thermal conductivity can only be calculated from the truncated HCACF.

Previous study suggested to determine cut-off time based on the first dip (FD) method when the tail of HCACF first decays to zero [56]. This corresponds to the time when the accumulative thermal conductivity, defined as

$$\kappa_a(t) = \frac{t_0}{3k_B T^2 V} \sum_{t=0}^{t} Cor(t), \tag{1.65}$$

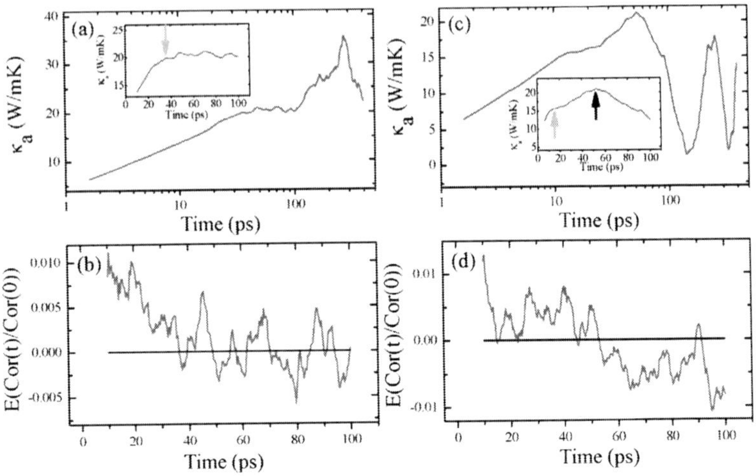

Figure 1.18 Accumulative thermal conductivity (curve in a and c) and mean value of normalized HCACF (curve in b and d) for the crystalline Silicon at 1000 K. Here we show two typical realizations in a 4 × 4 × 4 super cell. The inset zooms in for the short-time region. The light and dark arrows pinpoint the cut-off time estimated by first avalanche and first dip, respectively. The straight line (b,d) draws the zero-axis for reference.

first reaches a plateau (Fig. 1.18a). This is the ideal case that the average of HCACF fluctuates around zero in a relatively short time (Fig. 1.18b), and there is no accumulation of noise after the cut-off time. However, Chen *et al.* [65] found that FD method is not always reliable in practical EMD simulations where no obvious plateau can be found in HCACF, as shown in Fig. 1.18c. This corresponds to the realization that the mean value of HCACF can remain positive for a relatively long time even after HCACF decays to zero (Fig. 1.18d). In this case, FD method determines the cut-off time according to the peak of the accumulative thermal conductivity (black arrow in Fig. 1.18c), while there is already quite a lot of fluctuation in HCACF long before the cut-off time.

In order to get a quantitative description of the numerical error, Chen *et al.* [65] define the relative fluctuation of HCACF as

$$F(t) = \left| \frac{\sigma\,(Cor(t))}{E\,(Cor(t))} \right|, \tag{1.66}$$

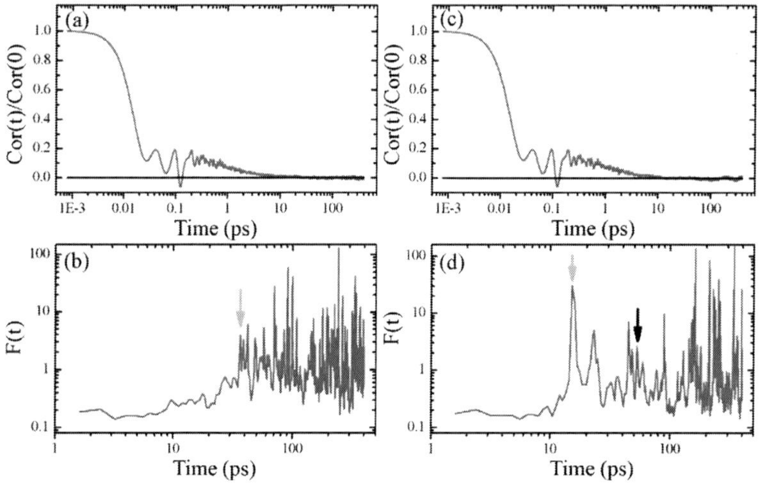

Figure 1.19 Time dependence of normalized HCACF Cor(t)/Cor(0) (curve in a and c) and its relative fluctuation F(t) (curve in b and d) for the same two realizations show in Fig. 1.18. The light and dark arrows pinpoint the cut-off time estimated by first avalanche and first dip, respectively. The straight line (a,c) draws the zero-axis for reference.

where σ and E denote the standard deviation and mean value of HCACF in the time interval $(t, t+\delta)$, respectively.

Figure 1.19a,c shows the time dependence of HCACF for the same two realizations shown in Fig. 1.18. HCACF exhibits the typical two-stage decaying characteristics of HCACF, and decays to zero in the long-time region. As shown in Fig. 1.19b,d, before a critical time, the relative fluctuation of HCACF $F(t)$ maintains a small value (e.g., less than 1) and does not change significantly. This indicates that HCACF is still reliable as it shows no large fluctuation. After the critical time, $F(t)$ suddenly jumps to a large value, and changes value drastically over time, which is a typical signature of the random noise. This indicates that HCACF has been contaminated and dominated by computational error, and thus is no longer reliable. Therefore, Chen *et al.* [65] define this critical time (when $F(t)$ becomes larger than 1) as the cut-off time, as marked by the green arrow in Fig. 1.18 and 1.19, and refer this way to estimate τ_c as "first avalanche" (FA) method. The essence of FA is to only take into account those contributions from HCACF, which is before τ_c and discard the rest part of HCACF as noise. It is worth mentioning that the tiny but non-zero noise can accumulate over time and change the value of accumulative thermal conductivity dramatically as shown in Fig. 1.18a,c.

When an obvious plateau in κ_a can be observed (shown in Fig. 1.18a), FD can make estimation of τ_c quite close to that estimated by FA. Consequently, both two methods can give estimation of thermal conductivity quite close to each other. In this case, the mean value of the normalized HCACF fluctuates around zero in a relatively short time (Fig. 1.18b). As a result, the noise does not accumulate over time, giving rise to a plateau in κ_a with small fluctuations. However, in the case without any obvious plateau in κ_a (Fig. 1.18c), FD estimates τ_c (according to the peak in κ_a) to be about 52 ps (marked by black arrow), while $F(t)$ shows that HCACF has already been contaminated by noise after about 15 ps. In this case, FD overestimates the thermal conductivity by falsely taking into account the contribution from the positive noise (see Fig. 1.18d).

1.6.4 *Some Applications*

Next using Silicon nanowire (SiNW) and Silicon nanotube (SiNT) as examples [68], we show how to calculate thermal conductivity of nanomaterials with equilibrium molecular dynamics (EMD) simulations. The atomic structure of NW is initially constructed from diamond-structured bulk silicon, with N_X, N_Y, and N_Z unit cells in the x, y, and z directions, respectively. Here we set longitudinal direction along x-axis, and atoms in the same layers means they have the same x-coordinate. The structure of the NWs is constructed by selecting all the Si atoms that fall within a virtual cage placed in bulk silicon while silicon atoms falling outside this virtual cage are removed. For SiNT, some central atoms in SiNWs are removed to create the SiNTs structures (Fig. 1.20). The cross-section of the hollow region is rectangular, its size is controlled by two parameters L_y and L_z, which means there are $(2L_y + 1)$ and $(2L_z + 1)$ layers of silicon atoms removed away in y and $z$directions, respectively. With adjustable L_y and L_z (from 1 to 6), the cross-section area of SiNTs varies from 4.72 nm^2 to 7.30 nm^2.

Figure 1.20 Cross-sectional view of SiNTs with central atoms (smaller circles) denoting the removed atoms.

In this simulation, SW potential is used to derive the force term. Numerically, a velocity Verlet algorithm is employed to integrate Newton's equations of motion, and each MD step is set as 0.8 fs. A cubic supercell of $N_X \times N_Y \times N_Z$ unit cells is used. A periodic boundary condition is applied in the x (longitudinal) direction, and a free boundary condition is applied in the other two directions, the atoms on the inner and outer surfaces of SiNTs and SiNWs. For each realization, all the atoms are initially placed at their equilibrium positions but have a random velocity according to Gaussian distribution. Canonical ensemble MD with Langevin heat reservoir first runs for 10^5 steps to equilibrate the whole system at a given temperature. Then microcanonical ensemble MD runs for another 3×10^6 steps (2.4 ns) and heat current is recorded at each step. After that, thermal conductivity is calculated according

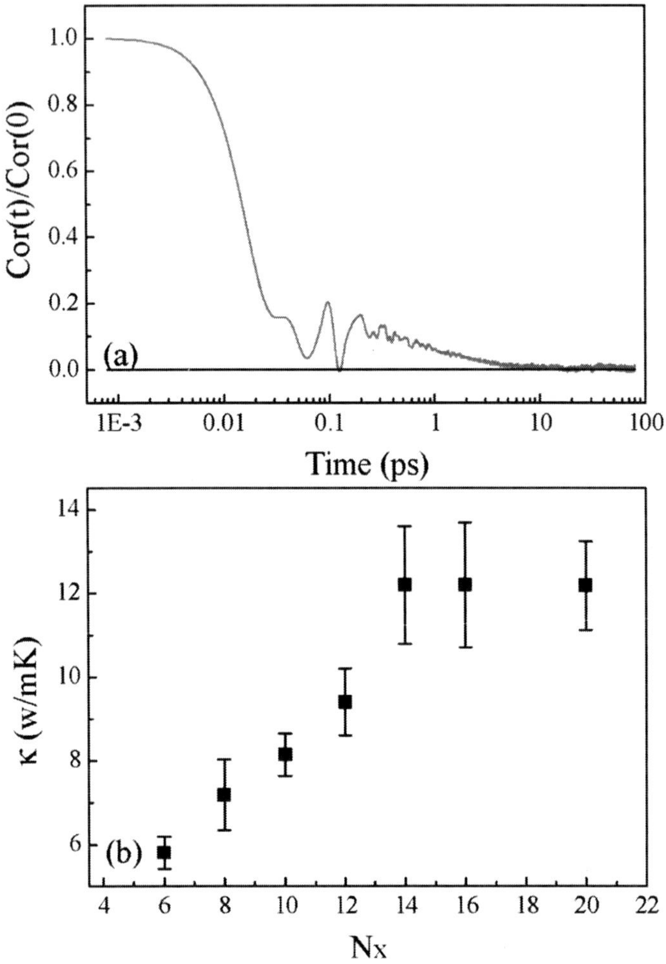

Figure 1.21 HCACF and thermal conductivity of SiNWs. (a) Time dependence of normalized HCACF *Cor(t)/Cor(0)* (red line) for a typical realization in a 16 × 5 × 5 supercell at 300 K. (b) Thermal conductivity of SiNWs at 300 K versus supercell size N_X ($N_X \times 5 \times 5$ unit cells).

to Eq. 1.47. The final result is averaged over six realizations with different initial conditions.

Figure 1.21a shows the time dependence of normalized HCACF for a typical realization in a 16 × 5 × 5 super cell at 300 K. It shows a very rapid decay of HCACF at the beginning, followed by a long tail which has a much slower decay. This two-stage decaying characteristic of HCACF has been found in the study of various materials. The rapid decay corresponds to the contribution from short-wavelength phonons to thermal conductivity, while the slower decay corresponds to the contribution from long wavelength phonons. Furthermore, as shown in Fig. 1.21a, HCACF decays to approximately zero within 100 ps (cut-off time), much shorter than the total simulation time of 2.4 ns. Therefore, the total simulation time of 2.4 ns is adequate for the present study.

Using the standard EMD approach in the integration of HCACF up to the cut-off time, the thermal conductivity of SiNWs was calculated. Figure 1.21b shows the calculated thermal conductivity of SiNWs with a fixed cross-section of 5×5 unit cells versus supercell length N_X at 300 K. Due to the periodic boundary condition, finite size effect exists in the calculated thermal conductivity when simulation domain is small. Here the thermal conductivity of SiNWs saturates to a constant when the supercell size is 16 unit cells. Therefore, in the following part, we set $N_X = 16$ in the longitudinal direction and study the thermal conductivity of SiNWs and SiNTs.

Figure 1.22 shows the thermal conductivity of SiNWs and SiNTs versus cross-section area at 300 K. Even with a very small hole, the thermal conductivity decreases evidently, from 12.2 W/(m-K) to 8.0 W/(m-K). In this case, only a 1% reduction in cross-section area induces the reduction of thermal conductivity of 35%.

Furthermore, with increasing of L_y and L_z, the cross-section area decreases further, and a linear dependence of thermal conductivity on cross-section area is observed. We also show the thermal conductivity of SiNW with $N_Y = N_Z = 4$. It has the same cross-section area (4.72 nm^2) as the SiNT with $N_Y = N_Z = 5$ and $L_y = L_z = 6$. It is clear that for SiNW, thermal conductivity increases as cross-section area increases. However, we can see that with the same cross-section area, thermal conductivity of SiNTs is only about 33% of that of SiNWs. This additional reduction is due to the localization of phonon states on the surface. Due to the inner surface in SiNTs, which partially destroys the original periodicity of SiNWs, phonon localization takes place on the surface in general.

In addition to the calculation of thermal conductivity, the HCACF can also provide more phonon information. The coherence of

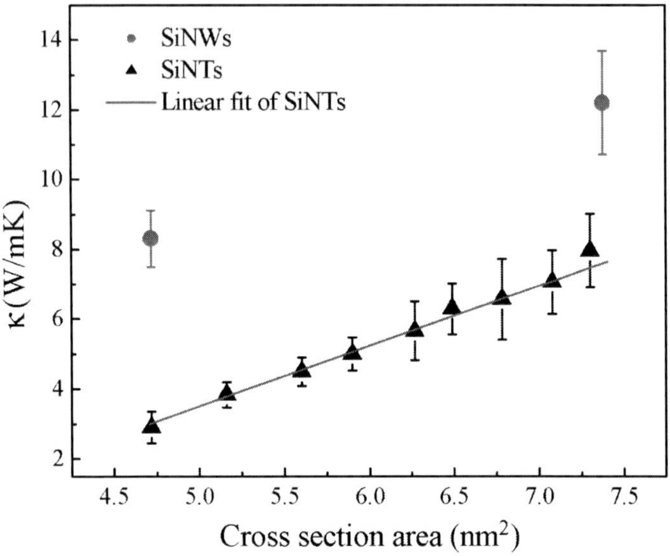

Figure 1.22 Thermal conductivity of SiNWs and SiNTs versus cross-section area at 300 K. The dot and triangle denote SiNWs and SiNTs, respectively. The cross-section areas for these two SiNWs are 7.37 nm^2 and 4.72 nm^2, respectively.

phonons has been probed by the HCACF in Si/Ge core–shell NWs in EMD simulations [69]. The schematic picture for core–shell NWs is shown in Fig. 1.23. The canonical ensemble molecular dynamics simulations with Langevin heat reservoir first run for 10^5 steps to equilibrate the NWs at a given temperature, during which the free boundary condition is applied to all the atoms on the surface.

After structure relaxation, all atoms are assigned with a random velocity according to Gaussian distribution. The periodic boundary condition is applied in the longitudinal direction, and the free boundary condition is applied in the other two directions. Then microcanonical ensemble EMD runs for 3×10^6 time steps (2.4 ns) and heat current is recorded at each step. Finally, the heat current (in longitudinal direction) autocorrelation function is calculated. The whole procedure is repeated six times with different initial conditions of the velocity distribution.

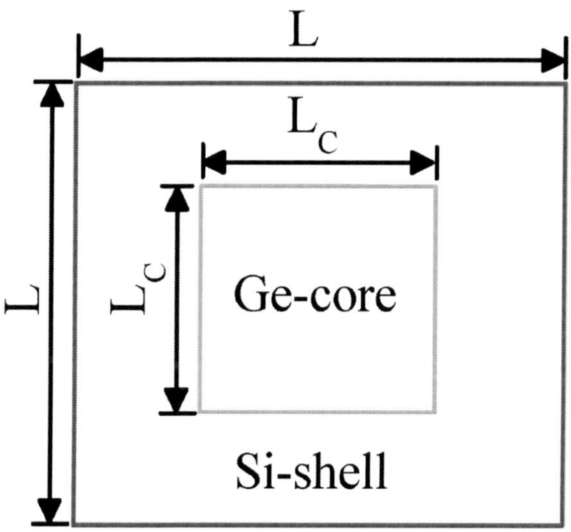

Figure 1.23 Cross-sectional view of Ge/Si core–shell NWs. The cross-sections of both core and shell regions are square, with L_c and L denoting the side length of core and shell regions, respectively.

Figure 1.24 shows the typical time dependence of normalized HCACF in a 16 × 5 × 5 super cell at 300 K for Ge/Si coreshell NWs with a given coreshell ratio ($L_c/L = 0.65$). For comparison, the HCACF for SiNWs and SiNTs is also shown. For both SiNWs and SiNTs, there is a very rapid decay of HCACF at the beginning, followed by a long-time tail with a much slower decay. When time

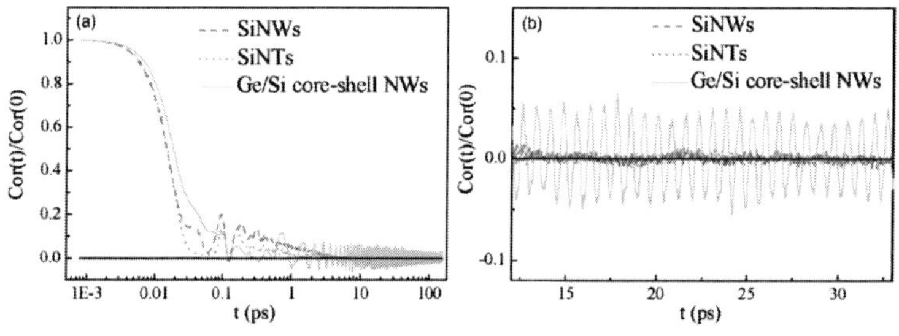

Figure 1.24 Time dependence of normalized HCACF. (a) Normalized HCACF for SiNWs, SiNTs, and Ge/Si core–shell NWs with $L_c/L = 0.65$. (b) Long-time region of (a).

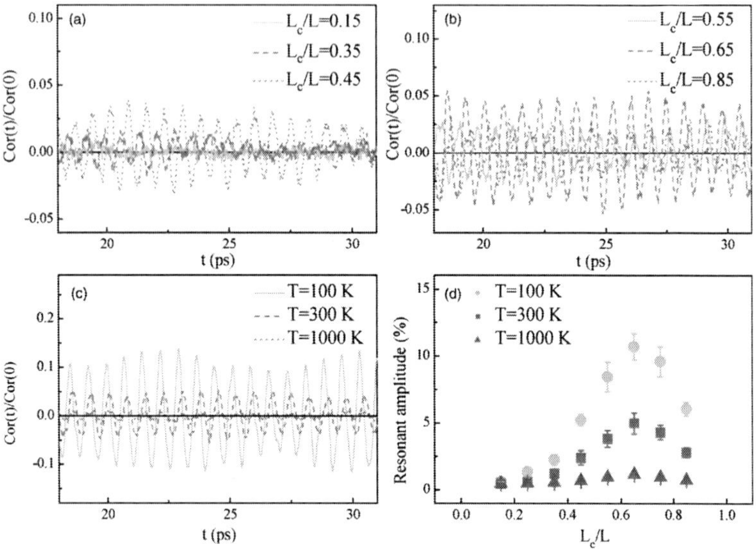

Figure 1.25 Structure and temperature dependence of the oscillation effect in Ge/Si core–shell NWs. (a) $L_c/L = 0.15$, $L_c/L = 0.35$, and $L_c/L = 0.45$ at 300 K. (b) $L_c/L = 0.55$, $L_c/L = 0.65$, and $L_c/L = 0.85$ at 300 K. (c) $L_c/L = 0.65$ at 100 K, 300 K, and 1000 K. (d) Oscillation amplitude versus core–shell ratio L_c/L at 100 K, 300 K, and 1000 K.

is long enough, the long-time tail of HCACF decays to approximately zero.

However, an obvious oscillation up to a very long time appears in HCACF for core–shell NWs. The long-time region of HCACF reveals that this oscillation has a periodic structure. Furthermore, this oscillation effect exhibits an obvious structure dependence: when the core–shell ratio increases, it becomes stronger, reaches its maximum amplitude at $L_c/L = 0.65$, and then decreases, as shown in Fig. 1.25a,b. Moreover, for a given core–shell structure, the oscillation amplitude is temperature dependent and becomes larger at lower temperature (Fig. 1.25c,d). The structure and temperature dependence of the oscillation in HCACF suggest that there exists a coherent mechanism in core–shell NWs, which can cause phonons to have the long-lasting correlation in such heterostructure.

Next, we carry out extensive spectrum analysis by using the fast Fourier transform (FFT). We calculate the FFT of the long-

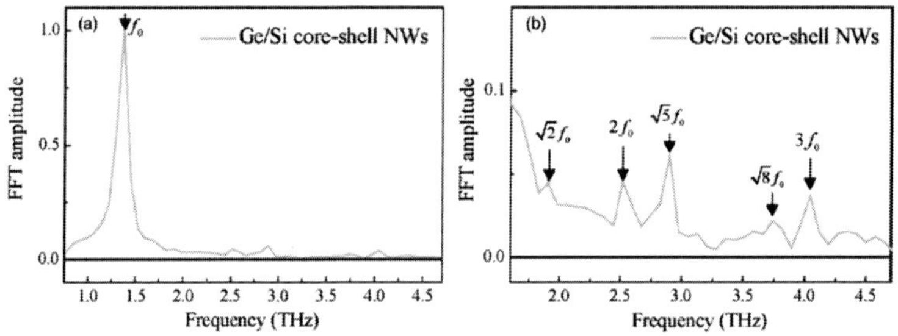

Figure 1.26 Amplitude of the FFT of the long-time region of normalized HCACF. The black lines in all figures draw the zero-axis for reference. (a) Ge/Si core–shell NWs. (b) The high-frequency oscillation peaks for Ge/Si core–shell NWs. The black arrows pinpoint the different oscillation frequencies.

time region of normalized HCACF. Figure 1.26 shows the FFT of normalized HCACF for different structures. The FFT amplitude for Ge/Si core–shell NWs exhibits a dominant peak at low frequency, which is denoted as f_0. All the FFT amplitudes shown here are normalized by the amplitude of the dominant peak. Moreover, there exist multiple high-frequency peaks, with much smaller amplitude compared to that of the dominant peak.

The multiple oscillation peaks observed in the frequency domain for core–shell NWs are very similar to the confinement effect of the acoustic wave (coherent long-wavelength phonon) in a confined structure. We record the high-frequency peaks marked by the black arrows in Fig. 1.26, and compare them with the frequency of the dominant peak f_0. We find that the relation between the frequencies of these high-frequency peaks and the dominant peak f_0 is very close to the eigenfrequency of the transverse modes for a wire with the square cross-section. This good agreement of oscillation frequency suggests that the intriguing oscillation effect results from the frequency quantization of the transverse modes as a consequence of structure confinement in the transverse direction.

In single-component homogeneous NWs, atoms on the same cross-section plane have the same sound velocity, so that the transverse motion is decoupled from the longitudinal motion. This is why the oscillation signal is not probed by the longitudinal HCACF in SiNWs. However, in core–shell NWs, atoms on the same cross-section plane have different sound velocity in the longitudinal direction. As a result, atoms near the core–shell interface are stretched due to the different sound velocity. This induces a strong coupling/interaction between the transverse and longitudinal motions. It is well known that when there is an interaction between two modes, the oscillation amplitude will be maximized when the frequencies of these two modes are close to each other (resonance). Due to the frequency quantization of the transverse modes, resonance will take place when the frequency of the longitudinal mode is close to the eigenfrequency of the transverse mode, giving rise to the enhanced oscillation amplitude. This coupling picture explains that frequency quantization of the transverse modes can indeed manifest itself in HCACF along the longitudinal direction in core–shell NWs. Moreover, as the resonance effect of acoustic wave is a coherent process that requires long-time correlation, the stronger anharmonic phonon–phonon scattering at high temperature causes phonon to lose coherence, and leads to the vanishing of the oscillation effect at high temperature. This study extends the application of EMD simulation, and provides a coherent resonance mechanism to tune thermal conductivity of nanomaterials.

References

1. Rahman, A. (1964) Correlations in the motion of atoms in liquid argon, *Phys. Rev.*, **136**, A405–A411.
2. Glassford, K. M., Chelikowsky, J. R., and Phillips, J. C. (1991) Interatomic force fields for the structure of intrinsic point defects in silicon, *Phys. Rev. B*, **43**, 14557–14563.
3. Ochoa, R., Swiler, T. P., and Simmons, J. H. (1991) Molecular dynamics studies of brittle failure in silica: effect of thermal vibrations, *J. Non-cryst. Solids*, **128**, 57–68.
4. Abraham, F. F. and Broughton, J. Q. (1986) Pulsed melting of silicon (111) and (100) surfaces simulated by molecular dynamics, *Phys. Rev. Lett.*, **56**, 734–737.
5. Landman, U., Luedtke, W. D., Barnett, R. N., Cleveland, C. L., Ribarsky, M. W., Arnold, E., Ramesh, S., Baumgart, H., Martinez, A., and Khan, B. (1986) Faceting at the silicon (100) crystal-melt interface: theory and experiment, *Phys. Rev. Lett.*, **56**, 155–158.

6. Feuston, B. P., Kalia, R. K., and Vashishta, P. (1987) Fragmentation of silicon microclusters: a molecular-dynamics study, *Phys. Rev. B*, **35**, 6222–6239.
7. Brenner, D. W., Dunlap, B. I., Harrison, J. A., Mintmire, J. W., Mowrey, R. C., Robertson, D. H., and White, C. T. (1991) Group-IV covalent clusters: Si_{45} and C_{44} versus Si_{44} and C_{45}, *Phys. Rev. B*, **44**, 3479–3482.
8. Berendsen, H. J. C. (1987) Biophysical applications of molecular dynamics, *Comp. Phys. Commun.*, **44**, 233–242.
9. Markussen, T., Jauho, A. -P., and Brandbyge, M. (2008) Heat conductance is strongly anisotropic for pristine silicon nanowires, *Nano Lett.*, **8**, 3771–3775.
10. Peelaers, H., Partoens, B., and Peeters, F. M. (2009) Phonon band structure of Si nanowires: a stability analysis, *Nano Lett.*, **9**, 107–111.
11. Yang, L., and Chou, M. Y. (2011) Lattice vibrational modes and their frequency shifts in semiconductor nanowires, *Nano Lett.*, **11**, 2618–2621.
12. Yang, N., Zhang, G., and Li, B. (2010) Violation of Fourier's law and anomalous heat diffusion in silicon nanowires, *Nano Today*, **5**, 85–90.
13. Chen, J., Zhang, G., and Li, B. (2011) A universal gauge for thermal conductivity of silicon nanowires with different cross sectional geometries, *J. Chem. Phys.*, **135**, 204705.
14. Hansen, J. P., and McDonald, I. R. (1986) *Theory of Simple Liquids*, 2nd Ed. (Academic Press).
15. Huang, K. (1987) *Statistical Mechanics*, 2nd Ed. (Jonh Wiley & Sons).
16. Turney, J. E., McGaughey, A. J. H., and Amon, C. H. (2009) Accessing the applicability of quantum corrections to classical thermal conductivity predictions, *Phys. Rev. B*, **79**, 224305.
17. Chen, J., Zhang, G., and Li, B. (2013) Substrate coupling suppresses size dependence of thermal conductivity in supported graphene, *Nanoscale*, DOI: 10.1039/C2NR32949B.
18. (a) Matyushov, D. V., and Schmid, R. (1996) Calculation of Lennard–Jones energies of molecular fluids, *J. Chem. Phys.*, **104**, 8627–8638. (b) Girifalco, L. A., Hodak, M., and Lee, R. S. (2000) Carbon nanotubes, buckyballs, ropes, and a universal graphitic potential, *Phys. Rev. B*, **62**, 13104–13110.
19. Rappé, A. K., Casewit, C. J., Colwell, K. S., Goddard III, W. A., and Skiff, W. M. (1992) UFF, a full periodic table force field for molecular mechanics and molecular dynamics simulations, *J. Am. Chem. Soc.*, **114**, 10024–10035.

20. Stillinger, F. H. and Weber, T. A. (1985) Computer simulation of local order in condensed phases of silicon, *Phys. Rev. B*, **31**, 5262–5271.
21. Tersoff, J. (1988) New empirical approach for the structure and energy of covalent systems, *Phys. Rev. B*, **37**, 6991–7000.
22. Tersoff, J. (1989) Modeling solid-state chemistry: interatomic potentials for multicomponent systems, *Phys. Rev. B*, **39**, 5566–5568(R).
23. Brenner, D. W. (1990) Empirical potential for hydrocarbons for use in simulating the chemical vapor deposition of diamond films, *Phys. Rev. B*, **42**, 9458–9471.
24. Brenner, D. W., Shenderova, O. A., Harrison, J. A., Stuart, S. J., Ni, B., and Sinnott, S. B. (2002) A second-generation reactive empirical bond order (REBO) potential energy expression for hydrocarbons, *J. Phys. Condens. Matter*, **14**, 783–802.
25. Nosé, S. (1984) A unified formulation of the constant temperature molecular dynamics methods, *J. Chem. Phys.*, **81**, 511.
26. Hoover, W. G. (1985) Canonical dynamics: equilibrium phase-space distributions, *Phys. Rev. A*, **31**, 1695–1697.
27. Dhar, A. (2008) Heat transport in low-dimensional systems, *Adv. Phys.*, **57**, 457–537.
28. Kubo, R. (1966) The fluctuation–dissipation theorem, *Rep. Prog. Phys.*, **29**, 255–284.
29. Holland, M. G. (1963) Analysis of lattice thermal conductivity, *Phys. Rev.*, **132**, 2461–2471.
30. Müller-Plathe, F. (1997) A simple nonequilibrium molecular dynamics method for calculating the thermal conductivity, *J. Chem. Phys.*, **106**, 6082.
31. Ikeshoji, T., and Hafskjold, B. (1994) Nonequilibrium molecular dynamics calculation of heat-conduction in liquid and through liquid-gas interface, *Mol. Phys.*, **81**, 251–261.
32. Chen, J., Zhang, G., and Li, B. (2010) Molecular dynamics simulations of heat conduction in nanostructures: Effect of heat bath, *J. Phys. Soc. Jpn.*, **79**, 074604.
33. Jiang, J.-W., Chen, J., Wang, J.-S., and Li, B. (2009) Edge states induce boundary temperature jump in molecular dynamics simulation of heat conduction, *Phys. Rev. B*, **80**, 052301.
34. Kim, S. Y., and Park, H. S. (2009) The importance of edge effects on the intrinsic loss mechanisms of graphene nanoresonators, *Nano Lett.*, **9**, 969–974.

35. Lima, M. P., Fazzio, A., and Silva, A. J. R. da (2009) Edge effects in bilayer graphene nanoribbons: ab initio total-energy density functional theory calculations, *Phys. Rev. B*, **79**, 153401.
36. Yang, L., Cohen, M. L., and Louie, S. G. (2008) Magnetic edge-state excitons in zigzag graphene nanoribbons, *Phys. Rev. Lett.*, **101**, 186401.
37. Bermudez, A., Patane, D., Amico, L., and Martin-Delgado, M. A. (2009) Topology-induced anomalous defect production by crossing a quantum critical point, *Phys. Rev. Lett.*, **102**, 135702.
38. Berendsen, H. J. C., Postma, J. P. M., van Gunsteren, W. F., DiHola, A., and Haak, J. R. (1984) Molecular dynamics with coupling to an external bath, *J. Chem. Phys.*, **81**, 3684.
39. Hunenberger, P. (2005) Thermostat algorithms for molecular dynamics simulations, *Adv. Polymer. Sci.*, **173**, 105–147.
40. Cheng, A. L., and Merz, K. M. (1996) Application of the Nose–Hoover chain algorithm to the study of protein dynamics, *J. Phys. Chem.*, **100**, 1927–1937.
41. Golo, V. L., and Shaitan, K. V. (2002) Dynamic attractor for Berendsen's thermostat and the slow dynamics, *Biofizika*, **47**, 611–617.
42. Mor, A., Ziv, G., and Levy, Y. (2008) Simulations of proteins with inhomogeneous degrees of freedom: the effect of thermostats, *J. Comput. Chem.*, **29**, 1992–1998.
43. Rosta, E., Buchete, N. V., and Hummer, G. (2009) Thermostat artifacts in replica exchange molecular dynamics simulations, *J. Chem. Theory Comput.*, **5**, 1393–1399.
44. Wu, G., and Li, B. (2007) Thermal rectification in carbon nanotube intramolecular junctions: molecular dynamics calculations, *Phys. Rev. B*, **76**, 085424.
45. Yang, N., Zhang, G., and Li, B. (2009) Thermal rectification in asymmetric graphene ribbons, *Appl. Phys. Lett.*, **95**, 033107.
46. Li, B., Wang, L., and Casati, G. (2004) Thermal diode: rectification of heat flux, *Phys. Rev. Lett.*, **93**, 184301.
47. Chang, C. W., Okawa, D., Majumdar, A., and Zettl, A. (2006) Solid-state thermal rectifier, *Science*, **314**, 1121–1124.
48. Zhang, G., and Li, B. (2005) Thermal conductivity of nanotubes revisited: effects of chirality, isotope impurity, tube length, and temperature, *J. Chem. Phys.*, **123**, 114714.
49. Chang, C. W., Okawa, D., Garcia, H., Majumdar, A., and Zettl, A. (2008) Breakdown of Fourier's law in nanotube thermal conductors, *Phys. Rev. Lett.*, **101**, 075903.

50. Chang, C. W., Fennimore, A. M., Afanasiev, A., Okawa, D., Ikuno, T., Garcia, H., Li, D., Majumdar, A., and Zettl, A. (2006) Isotope effect on thermal conductivity of Boron Nitride nanotubes, *Phys. Rev. Lett.*, **97**, 085901.
51. Chen, J., Zhang, G., and Li, B. (2012) Impacts of atomistic coating on thermal conductivity of Germanium nanowires, *Nano Lett.*, **12**, 2826–2832.
52. Green, M. S. (1954) Markoff random processes and the statistical mechanics of time-dependent phenomena. II. Irreversible processes in fluids, *J. Chem. Phys.*, **22**, 398–413.
53. Kubo, R. (1957) Statistical-mechanical theory of irreversible processes. I. General theory and simple applications to magnetic and conduction problems, *J. Phys. Soc. Jpn.*, **12**, 570–586.
54. Kubo, R., Yokota, M., and Nakajima, S. (1957) Statistical-mechanical theory of irreversible processes. II. Response to thermal disturbance, *J. Phys. Soc. Jpn.*, **12**, 1203–1211.
55. Lee, Y. H., Biswas, R., Soukoulis, C. M., Wang, C. Z., Chan, C. T., and Ho, K. M. (1991) Molecular-dynamics simulation of thermal conductivity in amorphous silicon, *Phys. Rev. B*, **43**, 6573–6580.
56. Li, J., Porter, L., and Yip, S. (1998) Atomistic modeling of finite-temperature properties of crystalline β-SiC II. Thermal conductivity and effects of point defects, *J. Nucl. Mater.*, **255**, 139–152.
57. Che, J., Cagin, T., Deng, W., and Goddard III, W. A. (2000) Thermal conductivity of diamond and related materials from molecular dynamics simulations, *J. Chem. Phys.*, **113**, 6888–6900.
58. Volz, S. G., and Chen, G. (2000) Molecular-dynamics simulation of thermal conductivity of silicon crystals, *Phys. Rev. B*, **61**, 2651–2656.
59. Zwanzig, R. (1965) Time-correlation functions and transport coefficients in statistical mechanics, *Ann. Rev. Phys. Chem.*, **16**, 67–102.
60. Kubo, R., Toda, M., and Hashitsume, N. (1985) *Statistical Physics*, Vol. II (Springer-Verlag: Berlin).
61. Kundu, A., Dhar, A., and Narayan, O. (2009) The Green–Kubo formula for heat conduction in open systems, *J. Stat. Mech. Theor. Exp.*, L03001.
62. Hardy, R. J. (1963) Energy-flux operator for a lattice, *Phys. Rev.*, **132**, 168–177.
63. Schelling, P. K., Phillpot, S. R., and Keblinski, P. (2002) Comparison of atomic-level simulation methods for computing thermal conductivity, *Phys. Rev. B*, **65**, 144306.

64. McGaughey, A. J. H., and Kaviany, M. (2004) Thermal conductivity decomposition and analysis using molecular dynamics simulations. Part. I. Lennard–Jones argon, *Int. J. Heat Mass Transfer*, **47**, 1783–1798.
65. Chen, J., Zhang, G., and Li, B. (2010) How to improve the accuracy of equilibrium molecular dynamics for computation of thermal conductivity? *Phys. Lett. A*, **374**, 2392–2396.
66. Glassbrenner, C. J., and Slack, G. A. (1964) Thermal conductivity of Silicon and Germanium from 3° K to the melting point, *Phys. Rev.*, **134**, A1058.
67. Couch, L. W. (2001) *Digital and Analog Communications Systems*, 6th Ed. (Prentice Hall).
68. Chen, J., Zhang, G., and Li, B. (2010) Remarkable reduction of thermal conductivity in silicon nanotubes, *Nano Lett.*, **10**, 3978–3983.
69. Chen, J., Zhang, G., and Li, B. (2011) Phonon coherent resonance and its effect on thermal transport in core–shell nanowires, *J. Chem. Phys.*, **135**, 104508.

Chapter 2

Non-equilibrium Phonon Green's Function Simulation and Its Application to Carbon Nanotubes

Takahiro Yamamoto,[a,b] Kenji Sasaoka,[c,d] and Satoshi Watanabe[c]

[a] *Department of Liberal Arts, Faculty of Engineering, Tokyo University of Science, 6-3-1 Niijuku, Katsushika-ku, Tokyo 125-8585, Japan*
[b] *Department of Electrical Engineering, Graduate School of Engineering, Tokyo University of Science, 6-3-1 Niijuku, Katsushika-ku, Tokyo 125-8585, Japan*
[c] *Department of Materials Engineering, Graduate School of Engineering, The University of Tokyo, 7-3-1 Hongo, Bunkyo-ku, Tokyo 113-8656, Japan*
[d] *Organization of Advanced Science and Technology, Kobe University, 1-1 Rokkodai-cho, Nada-ku, Kobe 657-8501, Japan*
watanabe@cello.t.u-tokyo.ac.jp

2.1 Introduction: Thermal Transport at Nanoscale

Theoretical analyses of thermal transport began from the pioneering work by Fourier in 1820s. According to his theory known as Fourier's law, the thermal current density I is proportional to the temperature gradient dT/dx, that is

$$I = -\lambda \frac{dT}{dx}, \tag{2.1}$$

Nanoscale Energy Transport and Harvesting: A Computational Study
Edited by Gang Zhang

ISBN 978-981-4463-02-7 (Hardcover), 978-981-4463-03-4 (eBook)
www.panstanford.com

where the thermal conductivity λ indicates the ability to conduct heat in materials. Fourier's law aptly describes the thermal-transport phenomena of a bulk material whose dimension is much larger than the mean free path and/or coherent length of heat carriers such as electrons and phonons. However, this law is no longer valid when the sample dimension reduces down to the mean free path of carriers. The problem of "breakdown of Fourier's law" originating from the reduction of sample size has been a long standing theoretical interest that goes back to Peierls' early work in the 1920s [1].

Over the past few decades, our understanding of phonon-derived thermal transport in such a small system has been advancing at a remarkable pace. Various striking phenomena that cannot be explained by Fourier's law have been discovered: the quantization of thermal conductance [2–5], the universal phonon-transmission fluctuation [6, 7], and the Anderson localization [8–10]. In this chapter, we present an overview of phonon transport or phonon-derived thermal transport in small objects. In Section 2.2, we give an account of a basic theory and simulation techniques of phonon-derived thermal transport that makes it possible to explain novel phonon transport phenomena of small objects with the breakdown of Fourier's law. In Section 2.3, we give a review of the non-Fourier thermal transport in carbon nanotubes as typical examples of actual nanomaterials.

2.2 Theory of Nanoscale Phonon Transport

In this section, we describe theory of thermal transport by coherent phonons in a quasi-one-dimensional (quasi-1D) system whose length is much shorter than the coherence length, which is bounded by phonon scattering events. We now introduce the Landauer approach for describing the coherent phonon transport in a quasi-1D system.

2.2.1 *Landauer Theory of Phonon Transport*

Let us consider the situation depicted in Fig. 2.1. A coherent phonon conductor in the absence of many-body interactions, such

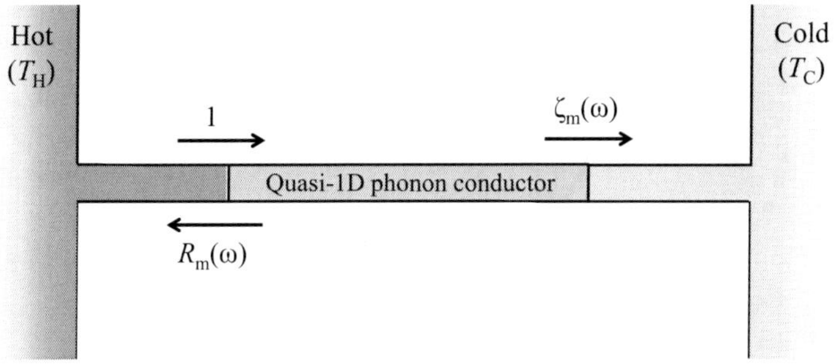

Figure 2.1 Landauer model for phonon transport in a quasi-1D conductor. The conductor is connected to two ideal leads.

as phonon–phonon and electron–phonon scatterings, is connected to ballistic quasi-1D leads without any scattering, which are in turn connected to heat reservoirs. The left and right heat reservoirs are assumed to be in thermal equilibrium with well defined temperatures T_H and T_C ($< T_H$), respectively. In this situation, the relaxation of phonons occurs only in the heat reservoirs. Thus, a phonon injected from the left and the right heat reservoirs follows the Bose–Einstein distribution function $f(\omega, T_H)$ and $f(\omega, T_C)$, respectively. Here, ω is the frequency of the injected phonon.

Let us derive a general expression of thermal current J and thermal conductance κ for the above situation. For simplicity, we consider a symmetric system having the same leads for the left and right regions. The thermal current carried by phonons with mode m and frequency ω injected from the left heat reservoir toward the central conductor is given by

$$\begin{aligned} j_m^L(\omega) &= \hbar\omega\,|v_m(\omega)|\,D_m^+(\omega)f(\omega, T_H) \\ &= \frac{1}{2\pi}\hbar\omega f(\omega, T_H) \end{aligned} \tag{2.2}$$

while the one injected from the right heat reservoir toward the central conductor is given by

$$\begin{aligned} j_m^R(\omega) &= \hbar\omega\,|v_m(\omega)|\,D_m^-(\omega)f(\omega, T_C) \\ &= \frac{1}{2\pi}\hbar\omega f(\omega, T_C) \end{aligned} \tag{2.3}$$

In the first lines in Eqs. 2.2 and 2.3, $D_m^{\pm}(\omega)$ denotes the phonon density of states for a unit length in the left and the right leads and the superscript $+$ $(-)$ stands for the density of states for phonons with positive (negative) group velocity $v_{\rm m}(\omega)$. Since $D^{\pm}_{\rm m}(\omega)$ is given by $D^{\pm}_{\rm m}(\omega) = D_{\rm m}(\omega)/2$ with $D_{\rm m}(\omega){=}1/(\pi\,|v_{\rm m}(\omega)|)$ for quasi-1D systems, it is canceled by the group velocity $v_{\rm m}(\omega)$ in the second line in Eqs. 2.2 and 2.3.

As shown in Fig. 2.1, $\zeta_{\rm m}(\omega)$ $[R_{\rm m}(\omega)]$ is the transmission (reflection) function for injected phonons with mode m and frequency ω. The net thermal current at the left lead, which is carried by phonons with m and ω, is thus given by

$$\begin{aligned} j_{\rm m}(\omega) &= j^{L}_{\rm m}(\omega)\,\{1 - R_{\rm m}(\omega)\} - j^{R}_{\rm m}(\omega)\zeta_{\rm m}(\omega) \\ &= \zeta_{\rm m}(\omega)\left\{j^{L}_{\rm m}(\omega) - j^{R}_{\rm m}(\omega)\right\} \\ &= \frac{1}{2\pi}\hbar\omega\zeta_{\rm m}(\omega)\,\{f(\omega, T_{\rm H}) - f(\omega, T_{\rm C})\} \end{aligned} \tag{2.4}$$

Similarly, we can obtain the same result as Eq. 2.4 for the net thermal current at the right lead. This can be interpreted as Kirchhoff's law for phonon transport. Using Eq. 2.4, the total thermal current, referred to as Landauer's energy flux, can be expressed as

$$\begin{aligned} J &= \sum_{m}\int_{\omega^{\rm min}_{\rm m}}^{\omega^{\rm max}_{\rm m}} j_{\rm m}(\omega)d\omega \\ &= \frac{1}{2\pi}\sum_{m}\int_{\omega^{\rm min}_{\rm m}}^{\omega^{\rm max}_{\rm m}} \hbar\omega\,[f_{\rm L}(\omega) - f_{\rm R}(\omega)]\,\zeta_{\rm m}(\omega)d\omega \end{aligned} \tag{2.5}$$

where $\omega^{\rm max}_{\rm m}$ and $\omega^{\rm min}_{\rm m}$ are the maximum and minimum values of frequency in the phonon dispersion curve for the mode m, respectively.

Let us consider the linear response situation with respect to the temperature difference $\Delta T = T_{\rm H} - T_{\rm C}$ between hot and cold heat reservoir. For a small temperature difference satisfying $\Delta T << (T_{\rm H} - T_{\rm C})/2$, the thermal current is given by

$$\begin{aligned} J &= \frac{\Delta T}{2\pi}\sum_{m}\int_{\omega^{\rm min}_{\rm m}}^{\omega^{\rm max}_{\rm m}} \hbar\omega\frac{df(\omega)}{dT}\zeta_{\rm m}(\omega)d\omega \\ &= \frac{\Delta T}{2\pi}\int_{0}^{\infty} \hbar\omega\frac{df(\omega)}{dT}\zeta(\omega)d\omega, \end{aligned} \tag{2.6}$$

where $\zeta(\omega) = \sum_{m}\zeta_{\rm m}(\omega)$ is the total transmission function [2–4]. Notice that the thermal current in Eq. 2.6 is not proportional to the temperature gradient dT/dx in contrast to Fourier's law in Eq. 2.2 for diffusive phonons, but is in fact proportional to the temperature difference ΔT. Using Eq. 2.6, we obtain the linear response form of thermal conductance as the following expression

$$\kappa(T) \equiv \frac{J}{\Delta T} = \frac{1}{2\pi}\int_0^{\infty} \hbar\omega \frac{df(\omega)}{dT}\zeta(\omega)d\omega \tag{2.7}$$

This is referred to as Landauer formula of phonon-derived thermal conductance [2–4]. The thermal conductance of quasi-1D systems can be estimated by calculating the phonon transmission function $\zeta(\omega)$ of the systems.

2.2.2 *Ballistic Phonon Transport and Quantization of Thermal Conductance*

According to the Landauer formula in Eq. 2.7, the thermal conductance is expressed by the phonon transmission function $\zeta(\omega)$. For the moment, we consider an ideal situation where all phonons are transmitted ballistically through a conductor without any scattering, namely $\zeta_m(\omega) = 1$ for all phonon modes. In this ballistic transport case, the total transmission function is given by

$$\zeta(\omega) = \sum_m \zeta_{\mathrm{m}}(\omega) = M(\omega), \tag{2.8}$$

where $M(\omega)$ is the number of phonon modes with angular frequency ω.

For quasi-1D systems such as nanowires, there are four acoustic modes, namely a longitudinal, a twisting, and two flexure acoustic modes (Fig. 2.2a). This is in a sharp contrast with usual 3D-bulk materials which have three acoustic modes: a longitudinal and two transverse acoustic modes. In the low-temperature region where the optical phonons are not excited, the thermal current flowing through a quasi-1D system is carried only by these four acoustic modes. Substituting $\zeta(\omega) = 4$ into Eq. 2.7, the low-temperature thermal conductance is quantized as

$$\kappa(T) = 4 \times \frac{k_{\mathrm{B}}^2 T}{h}\int_0^{\infty} \frac{x^2 e^{-x}}{(e^x - 1)^2}dx = 4\frac{\pi k_{\mathrm{B}}^2 T}{3h} \equiv 4\kappa_0, \tag{2.9}$$

where h is Planck's constant and

$$\kappa_0 = \frac{\pi k_{\mathrm{B}}^2 T}{3h} = (9.4 \times 10^{-13}\ \mathrm{W\ /\ K^2}) \times T \tag{2.10}$$

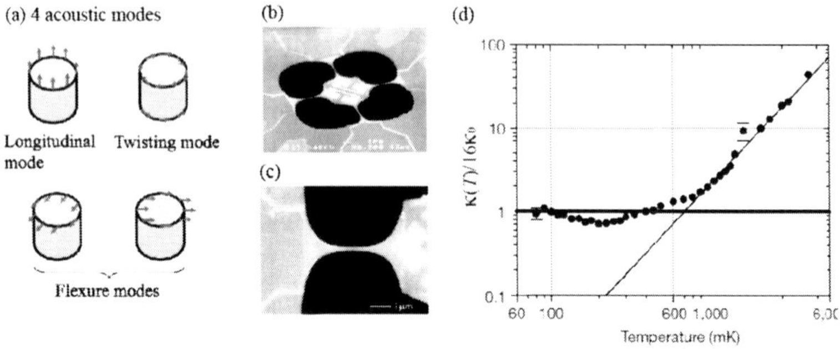

Figure 2.2 (a) The four acoustic modes of quasi-1D system. (b) The experimental setup used to measure the quantum of thermal conductance κ_0. (c) The enlarged view of a nanowire consisting of silicon nitride in Fig. 2.2a. (d) The temperature dependence of thermal conductance, normalized to a universal value of $16\kappa_0$, of the nanowire. Adapted by permission from Macmillan Publishers Ltd: Nature (Ref. [5]), copyright 2000.

is referred to as the thermal conductance quantum and gives the upper limit of the thermal conductance for a single conduction channel [2–4]. This fundamental upper limit coming from quantum mechanics can be regarded as a quantum of thermal conductance.

As seen in Eq. 2.9, the thermal conductance at low temperature in the ballistic limit is independent of the system length *L*. This is clearly different from the result derived by Fourier's law: the thermal conductance is inversely proportional to *L*. In this sense, the quantization of thermal conductance is one of the non-Fourier thermal-transport phenomena caused by the reduction of system size and is one of the most striking discoveries in the mesoscopic and nanoscale thermal transport.

The quantization of thermal conductance is very similar to the quantization of electrical conductance $G_0 = e^2/h$ in quantum wires. It is noted that the thermal conductance quantum κ_0 contains the thermal energy $k_B T$ instead of the electrical charge e in the electrical conductance quantum. This is because the thermal current carries the thermal energy $k_B T$ while the electrical current transports the charge e.

It is very challenging to observe the quantized thermal conductance κ_0 experimentally, because its value is extremely small as given in Eq. 2.10. In 2000, Schwab and his colleagues succeeded in detecting it in a nanowire consisting of silicon nitride below 0.6 K using a sophisticated fabrication technique [5]. Figure 2.2a shows the experimental setup used to measure the quantum κ_0. The small square at the center is a silicon nitride (SiN) membrane. The membrane is heated by two C-shaped heaters made by a gold thin film and acts as a heat reservoir. Since this membrane is suspended by four SiN bridges in vacuum, the heat reservoir and the four bridges are thermally isolated from the environment and the heat flows from the membrane into 4 SiN nanowire without heat dissipation. The perfect phonon transmission was realized by constructing smooth connections between the membrane and the SiN nanowires, as shown in Fig. 2.2b.

Figure 2.2c shows the temperature dependence of the phonon-derived thermal conductance of SiN nanowires normalized by $16\kappa_0$. The total thermal conductance of the device in Fig. 2.2a is expected to be $16\kappa_0$ because each of the four acoustic modes in each of the four bridges carries the quantum κ_0. At an extremely low temperature, below a threshold temperature of $\sim$0.6 K, the normalized thermal conductance is one or lower, as we expect. Above the threshold temperature, the thermal conductance curve increases because the optical phonon modes begin to contribute to the thermal transport.

As well as in SiN nanowires, quantization of the thermal conductance has also been observed in carbon nanotubes [11] just after its theoretical prediction [12].

2.2.3 *Non-equilibrium Green's Function Method for Phonon Transport*

The Landauer formula in Eq. 2.7 tells us that the thermal conductance is determined by the phonon transmission function $\zeta(\omega)$. Except for the ideal case of perfect transmission for all modes, it is difficult to calculate analytically the transmission $\zeta(\omega)$ for nanoscale objects with complex atomic structures. One of the powerful numerical techniques to calculate $\zeta(\omega)$ for nanomaterials is the non-equilibrium Green's function (NEGF) method [13–15]. According to the scattering theory using the NEGF technique, the total phonon transmission function is expressed as

$$\zeta(\omega) = Tr\left[\underline{\Gamma}_{R}(\omega)\underline{G}(\omega)\underline{\Gamma}_{L}(\omega)\underline{G}^{\dagger}(\omega)\right] \tag{2.11}$$

in terms of the retarded Green's function $\underline{G}(\omega)$ of the central conductor connected to the left and right leads. Here, the retarded Green's function is given by

$$\underline{G}(\omega) = \left[\omega^2 + i0_{+} - \underline{D} - \underline{\Sigma}_{L}(\omega) - \underline{\Sigma}_{R}(\omega)\right]^{-1}, \tag{2.12}$$

where the underlined quantities represent matrices with a basis in the central region, $\underline{D}$ is the dynamical matrix of the central region, which is derived from the second derivative of the total energy with respect to the atom coordinates in the central region, and $\underline{\Sigma}_{L/R}(\omega)$ is the self energy due to the left/right lead [13–15]. Its imaginary part relates to the level width function $\underline{\Gamma}_{L/R}$ as seen in Appendix A.

For the estimation of the phonon transmission function through Eq. 2.11, we have to know the retarded Green's function matrix, $\underline{G}(\omega)$, in Eq. 2.12. The direct numerical procedure to calculate Eq. 2.12 requires a higher computational cost as the length of the conductor becomes longer. However, Eq. 2.11 shows that not all elements of the matrix are needed in real space and only part of the elements, which relate phonon propagation between atoms in the left side and those in the right side in the conductor, are necessary. In the following, we shortly review the recursive Green's function technique that enables us to calculate these matrix elements efficiently [16]. This technique is based on the Dyson's equation that the retarded Green's function within the harmonic approximation satisfies. In the absence of many-body interactions such as the phonon–phonon interaction, we can obtain the perturbed Green's function exactly, as far as the solution of the equation converges. In this technique, the Green's function in Eq. 2.12 is obtained by a recursive procedure. At each step of the procedure, a perturbation is added and some blocks of the Green's function are exactly calculated based on the Dyson's equation. As a result, we can reduce the computational memory and get a higher computational speed.

First, the conductor is divided into layers appropriately, depending on its geometry (crystal structure and interatomic bonds). For simplicity, we consider the conductor having a square lattice structure, and only the adjacent atoms are assumed to interact with

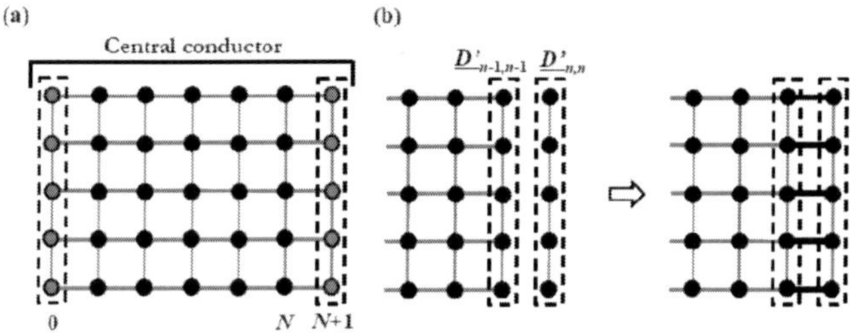

Figure 2.3 (a) A central conductor with a square lattice structure. The 0th and $(N + 1)$th columns are respectively the left and right buffer layers. Here N is the total number of layers of the scattering region in the central conductor. (b) nth iteration step in the recursive Green's function technique. nth layer is added into the subsystem constructed in the previous iteration.

each other, as shown in Fig. 2.3a. In this case, it is convenient to choose a column in the lattice as a layer. In addition, we suppose that the central conductor includes not only the scattering region but also portions of the left and right leads (0th and $(N + 1)$th layers), namely buffer layers. Here N is the number of the layers constituting the scattering region. Note that by introducing the buffer layers, all the other atoms in the leads can be regarded as the same as those in infinite leads, which makes the formulation easier. At this time, Green's functions in Eqs. 2.11 and 2.12, and the effective dynamical matrix of the central conductor can be represented by $(N + 2) \times (N + 2)$ block matrices. Then, the 0th and $(N + 1)$th intra-layer blocks of the effective dynamical matrix of the conductor $\underline{D}'$ are respectively described as

$$\underline{D}'_{0,0} = \underline{D}_{0,0} + \underline{\Sigma}'_{\mathrm{L}}(\omega), \quad \underline{D}'_{N+1,N+1} = \underline{D}_{N+1,N+1} + \underline{\Sigma}'_{\mathrm{R}}(\omega), \quad (2.13)$$

where $\underline{A}_{i,j}$ is a block element of a matrix $\underline{A}$ between the ith and jth layers. $\underline{D}$ denotes the dynamical matrix of the isolated conductor, and $\underline{\Sigma}'_{\mathrm{L/R}}(\omega)$is the intra-buffer-layer block of the self energy due to the left/right lead $\underline{\Sigma}_{\mathrm{L/R}}(\omega)$. All elements in the other blocks of the self energy matrices are zero, and so the other blocks of the effective dynamical satisfy the relation

$$\underline{D}'_{i,j} = \underline{D}_{i,j}. \quad (2.14)$$

It should be noted that the effective dynamical matrix has a block tri-diagonal structure with the diagonal blocks of $\underline{D}'_{i,i}$ and the off-diagonal blocks of $\underline{D}'_{i,i-1}$ and $\underline{D}'_{i-1,i}$.

Next, we treat all nearest neighboring inter-layer blocks of the dynamical matrix of the central conductor as a perturbation, while the intra-layer blocks are treated nonperturbatively. For this purpose, we consider the situation where these isolated layers are connected with the adjacent layers one by one from left to right of Fig. 2.3a: the 1st layer is added to the isolated left buffer layers (0th layer) as a perturbation. As mentioned above, this process is performed based on the Dyson's equation, which relates the perturbed Green's function matrix to the non-perturbed one. For the first step, that is, combining the 0th and 1st layers, we can obtain the following key relations:

$$\underline{G}_{1,1}^{(1)}(\omega) = \underline{g}_{1,1}(\omega) + \underline{g}_{1,1}(\omega)\underline{D}'_{1,0}\underline{G}_{0,1}^{(1)}(\omega), \tag{2.15}$$

$$\underline{G}_{0,1}^{(1)}(\omega) = \underline{g}_{0,0}(\omega)\underline{D}'_{0,1}\underline{G}_{1,1}^{(1)}(\omega), \tag{2.16}$$

$$\underline{G}_{1,0}^{(1)}(\omega) = \underline{G}_{1,1}^{(1)}(\omega)\underline{D}'_{1,0}\underline{g}_{0,0}(\omega), \tag{2.17}$$

where $\underline{G}_{i,j}^{(n)}(\omega)$ represents $(i,\ j)$ block of the Green's function on nth iteration step. The unperturbed Green's function is defined as

$$\underline{g}_{i,i}(\omega) = \left[\omega^2 + i0_+ - \underline{D}'_{i,i}\right]^{-1}. \tag{2.18}$$

Note that $\underline{G}_{1,1}^{(1)}(\omega)$ is self-consistently determined from Eqs. 2.15 and 2.16, and then $\underline{G}_{1,0}^{(1)}(\omega)$ can be obtained from Eq. 2.17. Next, the second layer is added to the combined layers, and we can obtain the following relations:

$$\underline{G}_{2,2}^{(2)}(\omega) = \underline{g}_{2,2}(\omega) + \underline{g}_{2,2}(\omega)\underline{D}'_{2,1}\underline{G}_{1,2}^{(2)}(\omega), \tag{2.19}$$

$$\underline{G}_{1,2}^{(2)}(\omega) = \underline{G}_{1,1}^{(1)}(\omega)\underline{D}'_{1,2}\underline{G}_{2,2}^{(2)}(\omega), \tag{2.20}$$

$$\underline{G}_{2,0}^{(2)}(\omega) = \underline{G}_{2,2}^{(2)}(\omega)\underline{D}'_{2,1}\underline{G}_{1,0}^{(1)}(\omega). \tag{2.21}$$

In the same manner as Eqs. 2.15 and 2.16, Eqs. 2.19 and 2.20 are solved self-consistently. The same procedure is repeated for the combined system and its adjacent isolated layer up to the $(N+1)$th iteration (see Fig. 2.3b), and eventually, the combined system becomes the same as the original system of Fig. 2.3a. Here, when $\underline{G}^{(0)}_{0,0}(\omega) = \underline{g_{0,0}}(\omega)$ is introduced, we can generalize the above relations for iterations as

$$\underline{G}^{(n)}_{n,n}(\omega) = \underline{g}_{n,n}(\omega) + \underline{g}_{n,n}(\omega)\underline{D}'_{n,n-1}\underline{G}^{(n)}_{n-1,n}(\omega), \tag{2.22}$$

$$\underline{G}^{(n)}_{n-1,n}(\omega) = \underline{G}^{(n-1)}_{n-1,n-1}(\omega)\underline{D}'_{n-1,n}\underline{G}^{(n)}_{n,n}(\omega), \tag{2.23}$$

$$\underline{G}^{(n)}_{n,0}(\omega) = \underline{G}^{(n)}_{n,n}(\omega)\underline{D}'_{n,n-1}\underline{G}^{(n-1)}_{n-1,0}(\omega). \tag{2.24}$$

Here, Eqs. 2.22 and 2.23 are also solved self-consistently. At the end of this iteration process, we can obtain $\underline{G}^{(N+1)}_{N+1,0}(\omega)$, which must be the block of the Green's function 2.12, $\underline{G}^{(N+1)}_{N+1,0}(\omega) = \underline{G}_{N+1,0}(\omega)$ because the system generated after $(N+1)$th iteration is equivalent to the original one. This equivalence is guaranteed since the Dyson's Eqs. 2.22, 2.23, and 2.24 are based on the exact perturbative treatments of the off-diagonal blocks of the dynamical matrix to the infinite order Now, the level width function is also constructed only by the intra-buffer-layer block and thereby $\underline{G}_{N+1,0}(\omega)$ is the block of the Green's function required for the calculation of the phonon transmission.

Next, we explain the method to compute the left/right surface Green's function $\underline{G}_{\mathrm{L}/\mathrm{R}}(\omega)$ [17], which is related to the self energy $\underline{\Sigma}_{\mathrm{L}/\mathrm{R}}(\omega)$ through

$$\underline{\Sigma}'_{\mathrm{L}/\mathrm{R}}(\omega) = \underline{D}^{+}_{\mathrm{LL}/\mathrm{RR}}\underline{G}_{\mathrm{L}/\mathrm{R}}(\omega)\underline{D}_{\mathrm{LL}/\mathrm{RR}}. \tag{2.25}$$

Here, $\underline{D}_{\mathrm{LL}/\mathrm{RR}}$ denotes the inter-layer dynamical matrix of the left/right semi-infinite lead. In addition, we denote the intra-layer matrix for a layer in the semi-infinite lead as $\underline{D}_{\mathrm{L}/\mathrm{R}}$. Note that the intra-layer dynamical matrix at the surface layer may be different from the ones in the bulk lead. So, we carry out the same procedure of the recursive technique depicted in Fig. 2.3b, that is, one layer is added to the semi-infinite lead in each iteration step. The semi-infinite system obtained after the iteration should be equivalent to the semi-infinite subsystem before the iteration. Therefore, by replacing $\underline{G}^{(n)}_{n,n}(\omega)$ and $\underline{G}^{(n-1)}_{n-1,n-1}(\omega)$ in Eqs. 2.22 and 2.23 with $\underline{G}_{\mathrm{L}/\mathrm{R}}(\omega)$, we can derive

$$\underline{G}_{\mathrm{L}/\mathrm{R}}(\omega) = \underline{g}_{\mathrm{L}/\mathrm{R}}(\omega) + \underline{g}_{\mathrm{L}/\mathrm{R}}(\omega)\underline{D}^{+}_{\mathrm{LL}/\mathrm{RR}}\underline{G}_{\mathrm{L}/\mathrm{R}}(\omega)\underline{D}_{\mathrm{LL}/\mathrm{RR}}\underline{G}_{\mathrm{L}/\mathrm{R}}(\omega) \tag{2.26}$$

where $\underline{g}_{L/R}(\omega)$ is defined as

$$\underline{g}_{L/R}(\omega) = \left[\omega^2 + i0_+ - \underline{D}_{L/R}\right]^{-1}. \tag{2.27}$$

The surface Green's function is self-consistently calculated by Eq. 2.26. As another method to obtain the self-energy due to the lead, the mode-matching method has been frequently used [18]. Both methods are originally developed in an electron system.

2.3 Application of Landauer–NEGF Method to Carbon Nanotube

A single-walled carbon nanotube (SWNT) depicted in Fig. 2.4, which is a single graphene layer rolled into a cylindrical shape, has a new quasi-1D crystalline form. Ever since the discovery of SWNTs in the early 1990s, it was anticipated that these nanostructures would have truly remarkable thermal-transport properties as well as electronic ones, given the strength of the carbon-carbon bond and the hexagonal network of carbons within a graphene layer. Nowadays, growing evidence, coming from both experimental and theoretical studies, indicates that SWNTs do indeed have extremely high thermal conductivities.

2.3.1 *Phonons in Carbon Nanotube*

Before discussing the thermal transport in carbon nanotube, we give a brief review of phonons in SWNTs. In general, the phonon dispersion relations of a system are obtained by diagonalizing the

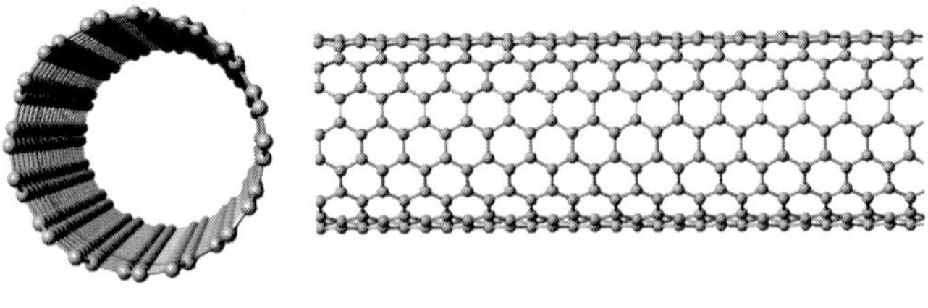

Figure 2.4 Schematics of atomic structure of a single-walled carbon nanotube.

dynamical matrix given by

$$D_{i\alpha}^{j\beta}(k) = \sum_{m} \frac{1}{\sqrt{M_i M_j}} K_{ni\alpha}^{mj\beta} \exp(ik \cdot (R_m - R_n)). \tag{2.28}$$

Here, the summation index m runs over all unit cells, $\boldsymbol{k}$ is the wave vector, $\boldsymbol{R}_m$ is the spatial coordinate of the mth unit cell, and M_i is the mass of the ith atom. $D_{i\alpha}^{j\beta}(k)$ is independent of n owing to the translational symmetry of the system. The force-constant $K_{ni\alpha}^{mj\beta}$ is defined as the second derivative of the potential with respect to atomic coordinates as follows.

$$K_{ni\alpha}^{mj\beta} = \frac{\partial^2 V}{\partial r_{ni\alpha} \partial r_{mj\beta}}, \tag{2.29}$$

where V is the potential energy of the system, $r_{ni\alpha}$ is the α component ($\alpha = x, y, z$) of the position vector of the ith atom in a unit cell n.

For carbon nanostructures, Brenner's empirical potential is adopted for V in Eq. 2.29. The Brenner potential has been successfully used for molecular-dynamic simulations of carbon nanostructures. Details of this potential can be found in Brenner's original paper [19]. Substituting the Brenner potential into Eq. 2.29, an analytical form of the force-constant can be obtained in a straightforward manner. The derivation is omitted here for the sake of brevity. As the force constants are only determined for the equilibrium positions of carbon atoms in the system, it is necessary to optimize the atomic geometry of the SWNTs using the Brenner potential.

Figure 2.5a shows phonon energy dispersion curves for the region near the Γ point ($\boldsymbol{k} = 0$) for a SWNT with chiral vector $\boldsymbol{C}_h =$ (10,10) [19, 20]. Here the chiral vector (n,m) uniquely determines the geometrical structure of SWNTs (See Appendix B for details of SWNT structure). Four acoustic modes are seen in Fig. 2.5, as mentioned previously, they are the longitudinal acoustic (LA) mode, the doubly degenerate flexure (FL) acoustic modes, and a twisting (TW) acoustic mode. The lowest optical (E_{2g} Raman active) modes are doubly degenerate. The excitation energy of E_{2g} Raman active modes depends only on the tube diameter d_t and decreases approximately according to $\sim 1/d_t^2$ [21]. These modes always lie in

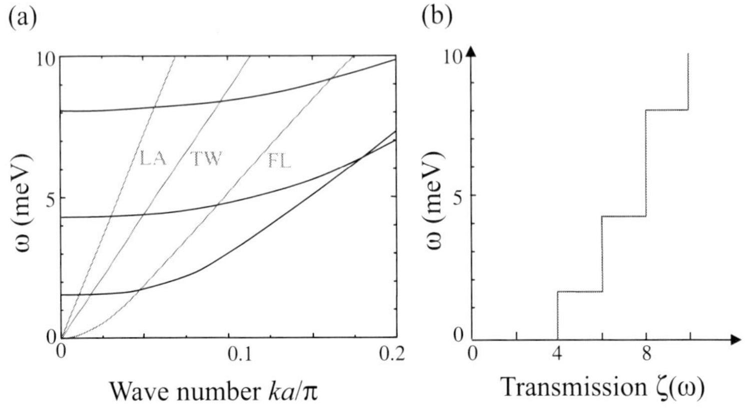

Figure 2.5 (a) The low-energy phonon dispersion relations. There are 4 acoustic modes: the longitudinal acoustic (LA), twisting (TW) modes, and two flexure modes. Here, a denotes the length of the unit cell in the (10, 10) single-walled carbon nanotube (SWNT). (b) The corresponding phonon transmission function of the (10,10) SWNT, respectively.

low-energy dispersion relations, independent of the tube geometry such as diameter or chirality.

In the case of the ballistic phonon transport, the phonon transmission function is given as a number of phonon dispersion branches, as shown in Eq. 2.8. In the low-temperature limit where the thermal energy $k_B T$ is much lower than the excitation energy of the lowest optical modes, the phonon-derived thermal conductance of SWNTs exhibits $4\kappa_0$ because there are 4 conduction channels as mentioned previously and seen in Fig. 2.5b [12]. In fact, the quantization of thermal conductance of nanotubes was observed experimentally [11].

2.3.2 *Thermal Conductance Reduction by Defect Scattering*

In this subsection, we apply the Landauer–NEGF formalism to the thermal transport in defective CNTs [13]. Figure 2.6 represents the phonon transmission function for the (8,8) SWNT with and without defects. The dashed curve is the transmission function $\zeta_p(\omega)$ for the perfect (8,8) SWNT without any defects, and it displays a clear stepwise structure that gives the number of phonon channels. In

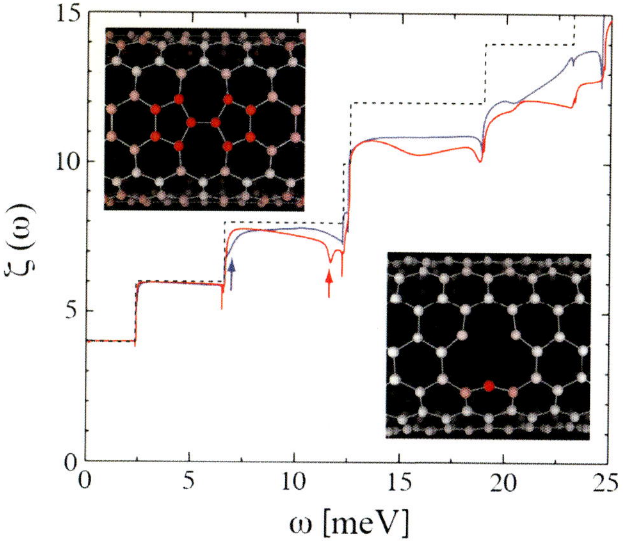

Figure 2.6 Phonon transmission function $\zeta(\omega)$ for the (8,8) SWNTs. Red and blue curves represent $\zeta(\omega)$ for SWNTs with vacancy and SW defect, respectively. The dashed curves are for the perfect SWNT. The top-left inset describes the phonon density around the SW defect at 7.0 meV indicated by the blue arrow, and the bottom-right inset is that around the vacancy at 11.6 meV indicated by the red arrow. The red shading on the atom spheres indicates the phonon density. (The red shading is not on the same scale for the two insets.)

the low-energy region below 2.4 meV, being the energy gap of the lowest optical modes, the dashed curve shows $\zeta_{\rm p}(\omega) = 4$ indicating the number of acoustic branches corresponding to the longitudinal, twisting, and doubly degenerated flexural modes.

The transmission function $\zeta_{\rm vac}(\omega)$ for the (8,8) SWNT with the vacancy and $\zeta_{\rm SW}(\omega)$ for the (8,8) SWNT with the Stone–Wales (SW) defects [22] are described by the red and blue curves, respectively, in Fig. 2.6. The $\zeta_{\rm vac\,/\,SW}(\omega)$ is dramatically deformed from $\zeta_{\rm p}(\omega)$ owing to defect scattering particularly at high energies. However, it remains unchanged in the low-energy region. This is because the long wavelength acoustic phonons with low energies in the CNTs are not scattered by local defects such as vacancies and SW defects. This leads to the important conclusion that the thermal conductance of

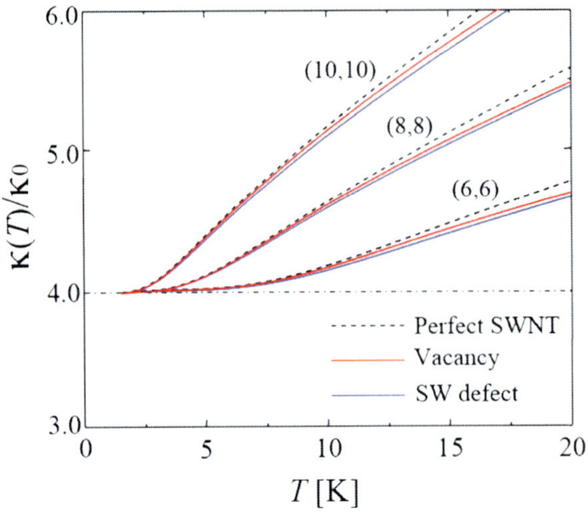

Figure 2.7 Low-temperature thermal conductances in (6,6), (8,8), and (10,10) SWNTs. Red and blue curves represent $\kappa(T)/\kappa_0$ for the SWNTs with the vacancy and SW defect, respectively. The dashed curves are for the perfect SWNT.

CNTs exhibits $4\kappa_0$ at cryostatic temperatures even when the CNTs include a small amount of defects (Fig. 2.7).

The transmission function $\zeta_{\text{vac}}(\omega)$ shows many dips. These dips can be classified into two groups by the origin of dip appearance. One is related to the van Hove singularity points where the density of states (DOS) of phonons diverges: these points are located at the top and bottom of phonon dispersion branches. Because the scattering rate is almost proportional to the DOS according to Fermi's golden rule, the transmission decreases around the van Hove singularity. The positions of the other group of dips, for example, the dip at 11.6 meV indicated by the red arrow, coincide with the peaks in the local density of states (LDOS) around the vacancy (not shown). The appearance of the LDOS peaks means that the phonon density is highly localized around the vacancy. The transmission dips arise from the scattering of incident phonons from the lead by the phonon localized states. The bottom-right inset in Fig. 2.6 shows the phonon density around the vacancy at 11.6 meV indicated by the red arrow in Fig. 2.6. The LDOS peak at 11.6 meV is located at the lowest position and has the largest intensity among the peaks. Similarly, $\zeta_{SW}(\omega)$ has some dips at particular energies coinciding with positions of the LDOS peak due to the SW defect.

The dip at 7.0 meV indicated by the blue arrow lies at the lowest position among the LDOS peaks associated with the SW defect. As shown in the top-left inset in Fig. 2.6, the phonon density at 7.0 meV is strongly localized around the SW defect.

Substituting the $\zeta(\omega)$ obtained into Eq. 2.7, we can determine the thermal conductance $\kappa(T)$ as a function of temperature T. The low-temperature behavior of $\kappa(T)$ normalized by the universal quantum κ_0 for (6,6), (8,8) and (10,10) SWNTs with and without defects are shown in Fig. 2.7. The red (blue) curves represent $\kappa_{vac(SW)}/\kappa_0$ for the CNTs with the vacancy (SW defect). The dashed curves represent κ_p/κ_0 for perfect CNTs. As discussed above, all thermal conductance curves approach 4 in the limit of $T \to 0$, even in the presence of local defects. In the temperature region shown in Fig. 2.7, the thermal conductance κ_{SW}/κ_0 is slightly lower than κ_{vac}/κ_0. This is because propagating phonons in this region are dominantly scattered by the phonon localized state associated with the SW defects at 7.0 meV, and not by the phonon localized states associated with a vacancy at the higher energy of 11.6 meV (Fig. 2.6).

We next describe the tube diameter dependence of defect scattering on thermal conductance for moderate temperatures up to 300 K. Figure 2.8 shows the ratios κ_{vac}/κ_p and κ_{SW}/κ_p for (6,6), (8,8), and (10,10) SWNT s as a function of T. The black, red, and blue solid (dashed) curves are the ratios $\kappa_{vac(SW)}/\kappa_p$ for (6,6), (8,8), and (10,10) SWNTs with the vacancy (SW defect), respectively. All the $\kappa_{vac(SW)}/\kappa_p$ curves decrease rapidly with increasing temperature and become nearly independent of the temperature at $\sim$300 K. The terminal value of $\kappa_{vac(SW)}/\kappa_p$ also decreases as the CNT gets thinner. In other words, the influence of defect scattering in *thin* CNTs on the thermal conductance is more significant than that in *thick* CNTs. Interestingly, κ_{vac}/κ_p is clearly lower than κ_{SW}/κ_p at moderate temperatures, which is in sharp contrast with the low temperature case. That is, the incident phonons are scattered more strongly by the vacancy than by the SW defect. This result is consistent with results obtained by classical molecular dynamics simulations at room temperature [23].

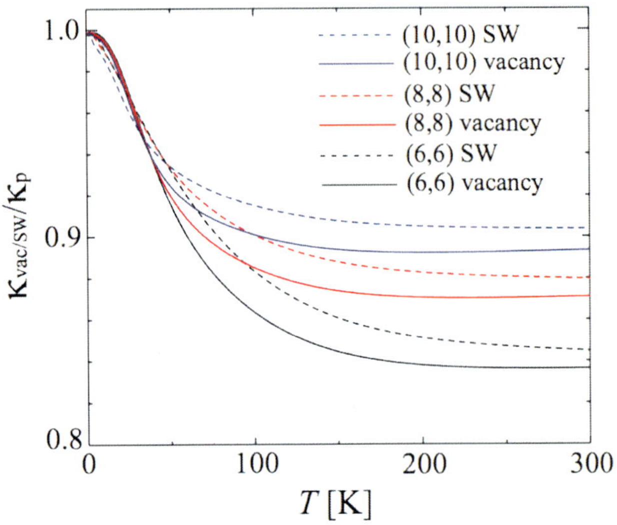

Figure 2.8 The temperature dependence of the ratio $\kappa_{\mathrm{vac(SW)}}/\kappa_p$ for the SWNT with the vacancy (SW defect). The black, red, and blue solid (dashed) curves are $\kappa_{\mathrm{vac(SW)}}/\kappa_{\mathrm{p}}$ for (6,6), (8,8), and (10,10) SWNTs, respectively.

2.3.3 *Isotope Effects on Thermal Transport in Carbon Nanotubes*

In this subsection, we present applications of Landauer–NEGF method to the coherent phonon transport in isotope-disordered SWNTs with randomly distributed ^{13}C and ^{14}C [24, 25]. Since the coexistence of ^{13}C and ^{14}C isotopes and the resultant elastic scattering cause the interference between coherent phonons, the isotope-disordered SWNTs is expected to exhibit remarkable phonon transport phenomena originating from the wave nature of phonons.

2.3.3.1 Characteristic lengths: mean free path and localization length

Coherent-phonon transport is classified into three regions based on a relation among the length L of the scattering region, the mean free path $l_{\mathrm{MFP}}(\omega)$, and the localization length ξ: the ballistic regime for $L \ll l_{\mathrm{MFP}}(\omega)$, the diffusive regime for $l_{\mathrm{MFP}}(\omega) \ll L \ll \xi(\omega)$, and the

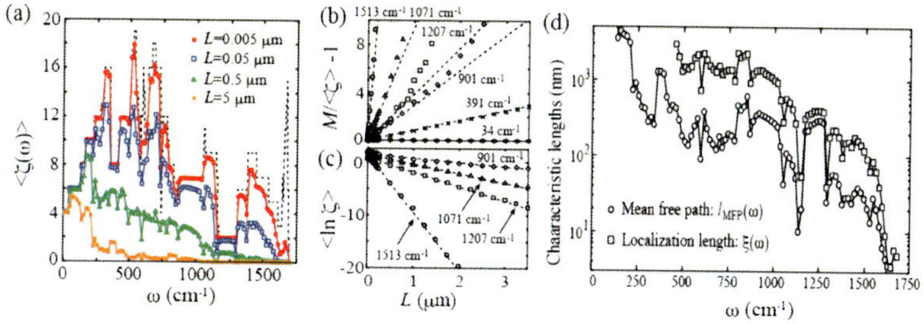

Figure 2.9 (a) The average phonon transmission $<\zeta(\omega)>$ of the (5,5) SWNT with 15% ^{13}C for $L = 0.005$ μm (circles), 0.05 μm (rectangles), 0.5 μm (triangles), and 5 μm (crosses). Averaging was performed over 200 random configurations. (b) The L-dependence of $M(\omega)/<\zeta(\omega)> -1$ for several frequencies, where $M(\omega)$ is the number of phonon modes of pristine SWNT. The dashed lines indicate linear fits for estimating the mean free path. (c) The L-dependence of $<\ln\zeta>$ for several frequencies. The dashed lines represent linear fits for estimating the localization length. (d) The mean free path (circles) and the localization length (rectangles) as functions of ω.

localization regime for $L \ll \xi(\omega)$ [24, 25]. Thus, we first determine $l_{\text{MFP}}(\omega)$ and $\xi(\omega)$ for isotope-disordered SWNTs as follows.

Figure 2.9a shows the average phonon transmission $<\zeta(\omega)>$ of the (5,5) SWNT with 15% ^{13}C for various L up to 5 μm. In the very low-frequency region, $<\zeta(\omega)>$ does not decrease and is almost four, even when isotope impurities are present. Perfect transmission (i.e., ballistic transport) is realized because the wavelength of acoustic phonons in the low-ω region is much longer than the length L. The Landauer expression of thermal conductance eventually exhibits universal quantization of $4\kappa_0$ at low temperatures even in the presence of isotope impurities (4 corresponds to the number of acoustic phonon modes).

In contrast, $<\zeta(\omega)>$ decreases rapidly in the higher frequency region with increasing L, as seen in Fig. 2.9a. There are two possible mechanisms for the reduction of $<\zeta(\omega)>$: diffusive scattering and Anderson localization. For diffusive scattering, $<\zeta(\omega)>$ decreases with L according to

$$\langle\zeta(\omega)\rangle = \frac{M(\omega)}{1 + L/l_{\text{MFP}}(\omega)}, \tag{2.30}$$

where $M(\omega)$ is the number of phonon modes with ω [27]. On the other hand, for phonon localization, the phonon-transmission function decays exponentially with L according to the scaling law [27]:

$$\langle \ln \zeta(\omega) \rangle = -\frac{L}{\xi(\omega)}. \tag{2.31}$$

In other words, the localization length $\xi(\omega)$ is defined by the scaling law. To clarify the mechanism for the phonon-transmission reduction, the L-dependences of phonon transmission are plotted in Fig. 2.9b and c for the two mechanisms, respectively. As Fig. 2.9b shows, the numerical data of $<\zeta(\omega)>$ at $\omega = 34$ cm^{-1} and 391 cm^{-1} are well fitted by the dashed lines. In particular, the slope of the dashed line for $\omega = 34$ cm^{-1} is almost zero, implying that the mean free path $l_{MFP}(\omega)$ is very long and the phonon transport is ballistic at this frequency, as has been discussed above. For $\omega = 391$ cm^{-1}, the slope is finite, which indicates that phonon transport at this frequency is in the diffusive regime. In contrast, at higher frequencies ($\omega = 1071$, 1207, and 1513 cm^{-1}), the calculated values deviate from the dashed lines with increasing L, although they are well fitted in the short-L region. This deviation means that the phonon-transmission reduction for high-ω phonons of a long-L SWNT cannot be explained by the diffusive scattering mechanism. As seen in Fig. 2.9c, the data for $\omega = 1071$, 1207, and 1513 cm^{-1} are well fitted by the dashed lines in the $<\zeta(\omega)>$ plot. We can thus conclude that the phonon-transmission reduction for high-ω phonons in a long-L SWNT is caused by Anderson localization of phonons.

The mean free path $l_{MFP}(\omega)$ and the localization length $\xi(\omega)$ can be estimated from the slope of dashed lines in Figs. 2.9b and c, respectively. The estimated $l_{MFP}(\omega)$ and $\xi(\omega)$ for the (5,5) SWNT with 15% ^{13}C are presented in Fig. 2.9d. Thus, the three distinct regimes (ballistic, diffusive, and localization) could be clarified. Similar results have been obtained for the (8,0) semiconducting SWNT with 9.4% ^{14}C. In the following subsections (Section 2.3.3.2 and Section 2.3.3.3), we discuss the phonon transport properties of the diffusive and the Anderson localization regimes, respectively.

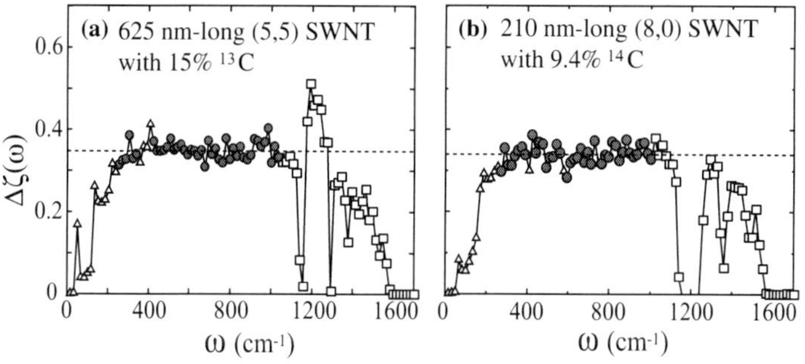

Figure 2.10 The root-mean-square phonon transmission for (a) a 625 nm-long (5,5) SWNT with 15% ^{13}C, and (b) a 210 nm-long (8,0) SWNT with 9.4% ^{14}C. The triangles, circles, and rectangles represent numerical data for $\Delta\zeta(\omega)$ in the ballistic, diffusive, and localization regimes, respectively. The dashed lines are the universal phonon-transmission fluctuation in Eq. 2.33.

2.3.3.2 Universal phonon-transmission fluctuation

We now discuss the phonon-transmission fluctuation, which is defined by a standard deviation

$$\Delta\zeta(\omega) = \sqrt{\langle\zeta(\omega)^2\rangle - \langle\zeta(\omega)\rangle^2}. \tag{2.32}$$

Figure 2.10 shows $\Delta\zeta(\omega)$ for (a) 625 nm-long (5,5) SWNT with 15% ^{13}C and (b) 210 nm-long (8,0) SWNT with 9.4% ^{14}C. The fluctuation of a physical quantity generally decreases as its average value increases. However, the fluctuation of phonon transmission is constant within the frequency region in the diffusive regime although $<\zeta(\omega)>$ varies depending on ω (Fig. 2.9a). The constant value is estimated to be

$$\Delta\zeta(\omega) = 0.35 \pm 0.02 \tag{2.33}$$

and is indicated by the dashed lines in Fig. 2.10a,b. Thus, $\Delta\zeta(\omega)$ in the diffusive regime is universal and does not depend on the background phonon transmission, the tube chirality and length, the isotope concentration, and differences between ^{13}C and ^{14}C. The universal phonon-transmission fluctuation is realized only in the diffusive regime and not in the ballistic and localization regimes. Interestingly, the value in Eq. 2.33 is the same as the value of the universal conductance fluctuation (UCF) for coherent electron

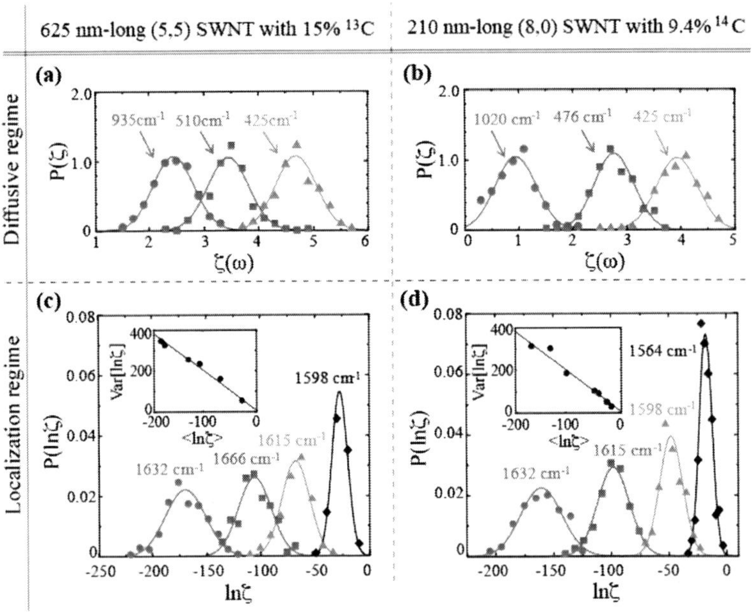

Figure 2.11 Phonon-transmission histograms $P(\zeta)$ for several frequencies in the diffusive regime for (a) the (5,5) SWNT with 15% ^{13}C and $L = 625$ nm and (b) the (8,0) SWNT with 9.4% ^{14}C and $L = 210$ nm. $P(\ln\zeta)$ for several frequencies in the localization regime for the same SWNTs as (c) Fig. 2.10a and (d) Fig. 2.10b. The insets show Var[ln ζ] as a function of $<P(\ln\zeta)>$.

transport in disordered quasi-1D systems, $\Delta G/G_0 = 0.36527$. This suggests that the universal phonon-transmission fluctuation is closely related to the UCF even though electrons and phonons obey different quantum statistics. Similar to the UCF, the reason for the macroscopically observable phonon transmission fluctuation can be qualitatively understood as follows: The fluctuations of phonon-transmission channels cannot cancel each other because there are very few effective transmission channels due to isotope scattering.

2.3.3.3 Anderson localization of phonons

In this subsection, we discuss the phonon-transmission histogram $P(\zeta)$ that contains information for every moment of $\zeta(\omega)$. Figure 2.11a,b show $P(\zeta)$ for several typical frequencies in the diffusive regime of (a) 625 nm-long (5,5) SWNT with 15% ^{13}C and (b) 210 nm-long (8,0) SWNT with 9.4% ^{14}C. All the histograms in Fig. 2.11a,b are well described by a Gaussian distribution function with the universal fluctuation $\Delta\zeta(\omega) = 0.35 \pm 0.02$ in Eq. 2.33. This is similar to the fact that the electrical conductance histogram in the diffusive region is expressed by a Gaussian distribution function with the UCF [26]. In the localization regime, $P(\zeta)$ becomes an asymmetric non-Gaussian distribution with a long tail towards large $\zeta(\omega)$. By analogy with the electrical conductance histogram in the localization regime [27, 28], the asymmetric histogram is expected to be a lognormal function of ζ. In fact, $P(\ln\zeta)$ can be well described by a Gaussian distribution, as seen in Fig. 2.11c, d. Unlike the diffusive regime, the variance $\mathrm{Var}[\ln\zeta] \equiv (\Delta \ln\zeta)^2$ of $P(\ln\zeta)$ decreases with increasing $<\ln\zeta>$ according to $\mathrm{Var}[\ln\zeta] = -2\langle\ln\zeta\rangle$ as seen in the insets of Fig. 2.11c,d, which is similar to the situation for electrons [27]. The phonon-transmission fluctuation in the localization regime is material independent in the sense that the slope of $\mathrm{Var}[\ln\zeta]$ does not depend on the tube geometry, isotope concentration, or the difference between ^{13}C and ^{14}C.

2.4 Concluding Remarks

Here, we first summarize the advantages of NEGF method as follows: (i) It is possible to calculate the fully-quantum phonon transport by NEGF method, while it is difficult to do that by non-equilibrium molecular dynamics methods. (ii) Because the NEGF method is excellent with the recursive algorithm as explained in Section 2.2.3, the phonon transport in large system, which is comparable to realistic materials, can be computed by NEGF method. (iii) The many-body interactions, such as phonon–phonon interaction, can be systematically incorporated into the NEGF formalism [29].

In this chapter, as applications of NEGF method, we review the influence of various defects on thermal-transport properties of carbon nanotube. We show that the carbon nanotube exhibits ballistic phonon transport at low temperature even in the presence of defects and eventually the thermal conductance of carbon nanotubes are quantized as the universal value at low temperature.

On the other hand, at the room temperature, even a single defect influences the thermal transport in carbon nanotube. The influence of defect scattering in *thin* carbon nanotubes on the thermal conductance is more significant than that in *thick* carbon nanotubes. We also explained the influence of randomlly-distributed ^{13}C isotopes on the thermal transport in carbon nanotubes. Interestingly, the phonon-transmission fluctuation in the diffusive regime is always 0.35 ± 0.02, irrespective of the average phonon transmission, the tube chirality, and the concentrations and masses of isotopes. We also found that the histogram, which has a Gaussian distribution in the diffusive regime, has a lognormal distribution in the localization regime. These features share much in common with disordered electron systems despite the different statistics of electrons and phonons. Moreover, we would emphasize that the knowledges obtained here are applicable to various low-dimensional nanostructures other than carbon nanotubes.

Recently, novel thermal devices such as a thermal rectifier [30–33], thermal transistor [34], and phonon waveguide [35, 36] have been proposed. However, many characteristic features of such novel devices remain to be elucidated yet. To design practical devices, it is essential to gain an understanding of the underlying mechanisms. The authors hope that this chapter will serve as a useful guideline for students and researchers who enter into this new and challenging field.

Appendix A: NEGF Formulation of Phonon Transmission Function

As mentioned in Section 2.23, the phonon transmission function can be expressed as Eq. 2.11. Here, we give a derivation of Eq. 2.11 in detail. The system Hamiltonian H_{sys} is described as the sum of the harmonic term

$$H_{\mathrm{har}} = \sum_{\substack{i \in sys \\ \alpha = xyz}} \left[\frac{p_{i\alpha}^2(t)}{2M_i} + \sum_{\substack{j>i \\ \beta = xyz}} \frac{k_{i\alpha,j\beta}}{2} \left(s_{i\alpha}(t) - s_{j\beta}(t) \right)^2 \right] \tag{A.1}$$

and the anharmonic term H_{anh}. Here, $s_{i\alpha}(t)$ is an operator in the Heisenberg picture for atomic displacement from equilibrium along the α direction of the ith atom with mass M_i. $p_{i\alpha}(t)$ is the momentum operator conjugated to the displacement operator $s_{i\alpha}(t)$, and $k_{i\alpha,j\beta}$ represents the spring constant between the ith atom in the α direction and the jth atom in the β direction. The system Hamiltonian is assumed to be divided into five parts:

$$H_{\text{sys}} = H_{\text{L}} + H_{\text{LS}} + H_{\text{S}} + H_{\text{RS}} + H_{\text{R}} \tag{A.2}$$

Here, $H_{\text{L/R}}$ is the Hamiltonian for the left/right thermal lead, H_{S} is that for the scattering region, and $H_{\text{LS(RS)}}$ is the Hamiltonian for the coupling between the scattering region and the left (right) lead. The anharmonic term H_{anh} is assumed to exist only in the scattering term H_{S}. Different temperatures, T_{L} and T_{R} ($<T_{\text{L}}$), are assigned to the left and right regions of the system, respectively.

The thermal current flowing through the interface between the left lead and the scattering region can be calculated from the time evolution of the energy of the left lead:

$$J = -\left\langle \frac{dH_{\text{L}}}{dt} \right\rangle = \frac{i}{\hbar}\left\langle \left[H_{\text{L}}, H_{\text{sys}}\right] \right\rangle \tag{A.3}$$

where the brackets in Eq. A.3 denotes the non-equilibrium statistical average of physical observable. The thermal current J is rewritten as

$$J = -\lim_{t' \to t} \sum_{\substack{i \in L, j \in S \\ \alpha\beta = xyz}} \frac{k_{i\alpha,j\beta}}{2} \left[i\hbar \frac{d}{dt'} G^{<}_{i\alpha,j\beta}(t,t') + H.c. \right] \tag{A.4}$$

using the greater and lesser Green's functions associated with the contact between the left lead and the scattering region:

$$i\hbar G^{>}_{i\alpha,j\beta}(t,t') = \left\langle s_{i\alpha}(t), s_{j\beta}(t') \right\rangle \tag{A.5}$$

and

$$i\hbar G^{>}_{i\alpha,j\beta}(t,t') = \left\langle s_{j\beta}(t'), s_{j\alpha}(t) \right\rangle \tag{A.6}$$

Since the Green's functions depend only on the time difference in a steady state, it is convenient to work in Fourier space (ω space). Then, in the steady state, the thermal current is expressed as

$$J = -\sum_{\substack{i \in L, j \in S \\ \alpha\beta = xyz}} \int_0^{\infty} \frac{d\omega}{2\pi} \hbar\omega\, k_{i\alpha,j\beta} \left[G^{<}_{i\alpha,j\beta}(\omega) - G^{>}_{j\beta,i\alpha}(\omega) + H.c. \right] \tag{A.7}$$

where the relation $G^{<}_{i\alpha,j\beta}(-\omega) = G^{>}_{j\beta,i\alpha}(\omega)$ is used.

In accordance with the similar procedure in the NEGF formalism for electronic transport [37], Eq. A.7 can be straightforwardly rewritten as

$$J = \int_0^\infty \frac{d\omega}{2\pi} \hbar\omega \, Tr \left[\underline{\Sigma}_L^>(\omega) \underline{G}_S^<(\omega) - \underline{\Sigma}_L^<(\omega) \underline{G}_S^>(\omega) \right] \tag{A.8}$$

where the quantities with underlines represent matrices with basis in the scattering region. In Eq. A.8, $\underline{G}_S^{>,<}(\omega)$ is the greater/lesser Green's function for the scattering region, and $\underline{\Sigma}_{L,R}^{>,<}(\omega)$ is the greater/lesser self-energy due to coupling to the left/right lead, which is given by

$$\underline{\Sigma}_{\mathrm{L,R}}^{>,<}(\omega) = -i \left(f_{\mathrm{L,R}}(\omega) + \frac{1}{2} \pm \frac{1}{2} \right) \underline{\Gamma}_{\mathrm{L,R}}(\omega) \tag{A.9}$$

Here,

$$\underline{\Gamma}_{\mathrm{L,R}}(\omega) = i \left(\underline{\Sigma}_{\mathrm{L,R}}^r(\omega) - \underline{\Sigma}_{\mathrm{L,R}}^a(\omega) \right) \tag{A.10}$$

where $\underline{\Sigma}_{\mathrm{L,R}}^{r,a}(\omega)$ is the retarded/advanced self-energy due to the coupling to the left/right lead.

In the case of coherent phonon transport, the lesser and greater Green's functions $\underline{G}_S^{>,<}(\omega)$ satisfy the Keldysh equation:

$$\underline{G}_{\mathrm{S}}^{>,<}(\omega) = \underline{G}_{\mathrm{S}}^r(\omega) \left[\underline{\Sigma}_{\mathrm{L}}^{>,<}(\omega) + \underline{\Sigma}_{\mathrm{R}}^{>,<}(\omega) \right] \underline{G}_{\mathrm{S}}^a(\omega) \tag{A.11}$$

while the retarded and advanced Green's functions $\underline{G}_S^{r,a}(\omega)$ for the scattering region are expressed by

$$\underline{G}_{\mathrm{S}}^{r,a}(\omega) = \left[\omega^2 + i0_+ - \underline{D} - \left(\underline{\Sigma}_{\mathrm{L}}^{r,a}(\omega) + \underline{\Sigma}_{\mathrm{R}}^{r,a}(\omega) \right) \right]^{-1} \tag{A.12}$$

where $\underline{D}$ is the dynamical matrix derived from the second derivative of the total energy with respect to the atom coordinates in the scattering region.

Substituting Eqs. A.10 and A.11 into Eq. A.8, the thermal current in Eq. A.8 is reduced to the Landauer formula of Eq. 2.5 in the main text. The phonon transmission function $\zeta(\omega)$ in Eq. 2.5 is also expressed explicitly as

$$\zeta(\omega) = Tr \left[\underline{\Gamma}_{\mathrm{R}}(\omega) \underline{G}_{\mathrm{S}}^r(\omega) \underline{\Gamma}_{\mathrm{L}}(\omega) \underline{G}_{\mathrm{S}}^a(\omega) \right] \tag{A.13}$$

This expression of $\zeta(\omega)$ was derived by Mingo and Yang using an atomistic Green's function method [14] and most recently by Das and Dhar using Lippmann–Schwinger theory [38].

Appendix B: Structure of Single-Walled Carbon Nanotubes

Here, the structure of SWNTs is briefly summarized. Figure B.1 shows the graphene layer and the primitive lattice vectors $\mathbf{a}_1$ and $\mathbf{a}_2$, which are defined as

$$\mathbf{a}_1 = \left(\frac{\sqrt{3}a}{2}, \frac{a}{2}\right), \quad \mathbf{a}_2 = \left(\frac{\sqrt{3}a}{2}, -\frac{a}{2}\right) \tag{B.1}$$

Here, a is the Bravais lattice constant, $a = \sqrt{3}a_{cc} = 2.46\text{Å}$, where $a_{cc} = 1.42\text{Å}$ is the bond length between neighboring carbon atoms. As seen in Fig. B1, an SWNT can be conceptually obtained by folding the dashed line containing points O and B to the dashed line containing point A and B'. An SWNT structure is characterized by three geometrical parameters: the chiral vector $\mathbf{C}_\mathrm{h}$, the translation vector $\mathbf{T}$, and the chiral angle θ, as shown in Fig. B1.

The chiral vector $\mathbf{C}_\mathrm{h}$ or $\overrightarrow{OA}$ is expressed as

$$\mathbf{C}_h = n\mathbf{a}_1 + m\mathbf{a}_2 \tag{B.2}$$

that connects two crystallographically equivalent carbon atoms on the graphene. The SWNT structure is uniquely determined by the pair of integer (n,m) in Eq. B2. Thus, an SWNT is described as an (n,m) SWNT. For example, the circumference and diameter of an (n,m) SWNT are given by the absolute value of chiral vector $|\mathbf{C}_h| = a\sqrt{n^2 + m^2 + nm}$ and $d_t = |\mathbf{C}_h|/\pi = \left(a\sqrt{n^2 + m^2 + nm}\right)/\pi$, respectively. Other two geometrical parameters (T and θ) can be also determined from the chiral vector. The chiral angle θ is the angle between the chiral vector and the primitive lattice vector $\mathbf{a1}$:

$$\cos\theta = \frac{\mathbf{C}_h \cdot \mathbf{a}_1}{|\mathbf{C}_h|\,|\mathbf{a}_1|} = \frac{2n + m}{2\sqrt{n^2 + nm + m^2}} \tag{B.3}$$

Due to the sixfold hexagonal symmetry of graphene, the chiral angle is restricted to $0 \leq \theta \leq 30^\circ$. Next, the translational vector $\mathbf{T} = (t_1, t_2)$ can be determined by the orthogonality condition:

$$\mathbf{C}_\mathrm{h} \cdot \mathbf{T} = t_1(2n + m) + t_2(2m + n) = 0 \tag{B.4}$$

Thus, the integers t_1 and t_2 are given as

$$\mathbf{T} = (t_1, t_2) = \left(\frac{2m + n}{g_\mathrm{d}}, -\frac{2m + n}{g_\mathrm{d}}\right) \tag{B.5}$$

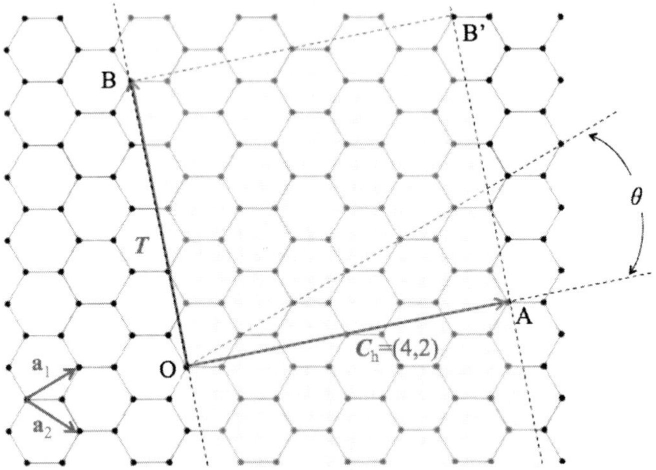

Figure B.1 Definition of chiral vector $\boldsymbol{C}_h$, translational vector $\boldsymbol{T}$, and chiral angle θ.

where $g_\mathrm{d} \equiv gcd(2m + n, 2n + m)$ is the greatest common divisor of $2m + n$ and $2n + m$. The length of the translational vector is

$$|\mathbf{T}| = T = \frac{\sqrt{3}\,|\mathbf{C}_\mathrm{h}|}{g_\mathrm{d}} \tag{B.6}$$

Further details of SWNT structure and others can be found in Ref. [21].

References

1. Peierls, R. E. (1955) *Quantum Theory of Solid* (Oxford University Press, New York).
2. Rego, L.G.C. and Kirczenow, G. (1998) Quantized thermal conductance of dielectric quantum wires, *Phys. Rev. Lett.*, **81**, 232–235.
3. Angelescu, D. E., Cross, M.C., and Roukes, M. L. (1998) Heat transport in mesoscopic systems, *Superlattices Microstruct.*, **23**, 673–689.
4. Blencowe, M. P. (1999) Quantum energy flow in mesoscopic dielectric structures, *Phys. Rev. B*, **59**, 4992–4998.

5. Schwab, K., Henrlksen, E. A., Worlock, J. M., and Roukes, M. L. (2000) Measurement of the quantum of thermal conductance, *Nature* (London), **404**, 974–977.
6. Nishiguchi, N., Tamura, S., and Franco, N. (1993) Phonon universal-transmission fluctuations and localization in semiconductor superlattices with a controlled degree of order, *Phys. Rev. B*, **43**, 14426–14435.
7. Yamamoto, T., Sasaoka, K., and Watanabe, S. (2011) Universality and diversity in a phonon-transmission histogram of isotope-disordered carbon nanotubes, *Phys. Rev. Lett.*, **106**, 215503.
8. Ishii, K. (1973) Localization of eigenstates and transport phenomena in the one-dimensional disordered system, *Prog. Theor. Phys. Suppl.*, **53**, 77–138.
9. Jackle, J. (1981) On phonon localization in nearly one-dimensional solids, *Solid State Commun.*, 39, 1261–1263.
10. Chaudhuri, A., Kundu, A., Roy, D., Dhar, A., Lebowitz, J. L. and Spohn, H. (2010) Heat transport and phonon localization in mass-disordered harmonic crystals, *Phys. Rev. B*, **81**, 064301.
11. Chiu, H.-Y., Deshpande, V. V., Postma, H. W. Ch., Lau, C. N., Miko, C., Forro, L., and Bockrath, M. (2005) Ballistic phonon thermal transport in multiwalled carbon nanotubes, *Phys. Rev. Lett.*, **95**, 226101.
12. Yamamoto, T., Watanabe, S., and Watanabe, K. (2004) Universal features of quantized thermal conductance of carbon nanotubes, *Phys. Rev. Lett.*, **92**, 075502.
13. Yamamoto, T. and Watanabe, K. (2006) Nonequilibrium Green's function approach to phonon transport in defective carbon nanotubes, *Phys. Rev. Lett.*, **96**, 255503.
14. Mingo, N. (2006) Anharmonic phonon flow through molecular-sized junctions, *Phys. Rev. B*, **74**, 125402.
15. Wang, J.-S., Wang, J., and Zeng, N. (2006) Nonequilibrium Green's function approach to mesoscopic thermal transport, *Phys. Rev. B*, **74**, 033408.
16. Thoulcss, D. J. and Kirkpatrick, S. (1981) Conductivity of the disorder linear chain, *J. Phys. C: Solid State Phys.*, **14**, 235–245.
17. Guinea, F., Tejedor, C., Flores, F., and Lous, E. (1983) Effective two-dimensional Hamiltonian at surfaces, *Phys. Rev. B*, **28**, 4397–4402.
18. Ando, T. (1991) Quantum point contacts in magnetic fields, *Phys. Rev. B*, **44**, 8017–8027.

19. Brenner, D. W. (1990) Empirical potential for hydrocarbons for use in simulating the chemical vapor deposition of diamond films, *Phys. Rev. B*, **42**, 9458–9471.
20. Yamamoto, T., Watanabe, K., and Mii, K. (2004) Empirical-potential study of phonon transport in graphitic ribbons, *Phys. Rev. B*, **70**, 245402.
21. Saito, R., Dresselhaus, G., and Dresselhaus, M. S. (1998) *Physical Properties of Carbon Nanotubes* (Imperical College Press, London).
22. Stone, A. J. and Wales, D. J. (1986) Theoretical studies of icosahedral C_{60} and some related species, *Chem. Phys. Lett.*, **128**, 501–503.
23. Kondo, N., Yamamoto, T., and Watanabe, K. (2006) Molecular-dynamics simulations of thermal transport in carbon nanotubes with structural defects, *e-J. Surf. Sci. Nanotech.*, **4**, 239–243.
24. Yamamoto, T., Sasaoka, K., and Watanabe, S. (2011) Universality and diversity in a phonon-transmission histogram of isotope-disordered carbon nanotubes, *Phys. Rev. Lett.*, **106**, 215503.
25. Savic, I., Mingo, N., and Stewart, D. A. (2008) Phonon transport in isotope-disordered carbon and boron-nitride nanotubes: Is localization observable?, *Phys. Rev. Lett.*, **101**, 165502.
26. Beenakker, C. W. (1997) Random-matrix theory of quantum transport, *Rev. Mod. Phys.*, **69**, 731–808.
27. Lee, P. A. and Stone, A. D. (1985) Universal conductance fluctuations in metals, *Phys. Rev. Lett.*, **55**, 1622–1625.
28. Imry, Y. (1986) Active transmission channels and universal conductance fluctuations, *Europhys. Lett.*, **1**, 249–256.
29. Wang, J.-S., Wang, J., and Zeng, N. (2006) Nonequilibrium Green's function approach to mesoscopic thermal transport, *Phys. Rev. B*, **74**, 033408.
30. Terraneo, M., Peyrard, M., and Casati, G. (2002) Controlling the energy flow in nonlinear lattices: a model for a thermal rectifier, *Phys. Rev. Lett.*, **88**, 094302.
31. Li, B., Wang, L., and Casati, G. (2004) Thermal diode: rectification of heat flux, *Phys. Rev. Lett.*, **93**, 184301.
32. Hu, B., Yang, L., and Zhang, Y. (2006) Asymmetric heat conduction in nonlinear lattices, *Phys. Rev. Lett.*, **97**, 124302.
33. Chang, C. W., Okawa, D., Majumdar, A., and Zettl, A. (2006) Solid-state thermal rectifier, *Science*, **314**, 1121–1124.
34. Li, B., Wang, L., and Casati, G. (2006) Negative differential thermal resistance and thermal transistor, *Appl. Phys. Lett.*, **88**, 143501.

35. Chang, C. W., Okawa, D., Garcia, H., Majumdar, A., and Zettl, A. (2008) Breakdown of Fourier's Law in nanotube thermal conductors, *Phys. Rev. Lett.*, **101**, 075903.

36. Nishimura, F., Takahashi, T., Watanabe, K., and Yamamoto, T. (2009) Bending robustness of thermal conductance of carbon nanotubes: Nonequilibrium molecular dynamics simulation, *Appl. Phys. Exp.*, **2**, 035003.

37. Haug, H. and Jauho, A.-P. (1996) *Quantum Kinetics in Transport and Optics of Semiconductors* (Springer, Berlin).

38. Das, S. G. and Dhar, A. (2012) Landauer formula for phonon heat conduction: Relation between energy transmittance and transmission coefficient, *Eur. Phys. J. B*, **85**, 372.

Chapter 3

Thermal Conduction of Graphene

Yong Xu[a,b] and Wenhui Duan[a]

[a] *Department of Physics, Tsinghua University, Beijing 100084, China*
[b] *Theory Department, Fritz-Haber-Institut der Max-Planck-Gesellschaft, Berlin 14195, Germany*
dwh@phys.tsinghua.edu.cn

3.1 Basic Concepts of Quantum Thermal Transport

3.1.1 *Thermal-Transport Carriers*

In physics, transport concerns the movements of physical entities (e.g., mass, charge, momentum, and energy) in the presence of inhomogeneities. Thermal transport, as one type of transport phenomena, is the exchange of energy (or heat) as a result of the temperature gradient. Electrons and phonons are major carriers of thermal transport in solids. In semiconductors and insulators, phonons, which are the collective excitation of lattice vibrations, are the main contributor to thermal transport. While in metals, a significant part of thermal conductance comes from electrons. A simple relation between the electronic contribution to the thermal conductivity (κ_e) and the electrical conductivity (σ_e) exists for metals. As revealed by the Wiedemann–Franz law, the ratio between

Nanoscale Energy Transport and Harvesting: A Computational Study
Edited by Gang Zhang

ISBN 978-981-4463-02-7 (Hardcover), 978-981-4463-03-4 (eBook)
www.panstanford.com

κ_e and σ_e is proportional to the temperature T

$$\frac{\kappa_e}{\sigma_e} = L_0 T. \tag{3.1}$$

L_0, known as the Lorenz number of free electrons, is equal to

$$L_0 = \frac{K_e}{\sigma_e T} = \frac{\pi^2 k_B^2}{3e^2} = 2.44 \times 10^{-8} \mathrm{W\Omega K^{-2}}. \tag{3.2}$$

When the size of materials decreases to nanoscale, thermal conduction is mainly contributed by phonons. For instance, previous experiments found that thermal conductivity of carbon nanotubes (CNTs) (including both metallic and semiconducting ones) is dominated by phonons at all temperatures [1]. This is partially because in nano-materials, confinement effects induce energy splitting of both electrons and phonons. The energy splitting is of the order of eV for electrons, while it is much smaller, of the order of meV, for phonons. Consequently, phonons are much easier to be excited by temperature than electrons, thus dominate thermal conduction in nano-materials. In the following sections, we focus on thermal conduction contributed by phonons.

3.1.2 *Fundamental Length Scales of Thermal Transport*

Thermal transport (i.e., phonon transport) has two fundamental length scales [2].

3.1.2.1 The characteristic wavelength of phonon λ

Assuming a Debye temperature of θ_D and a lattice constant of a; at high temperatures ($T > \theta_D$), $\lambda \sim a$; and at low temperatures ($T < \theta_D$), $\lambda \sim \theta_D a / T$. As the temperature decreases, λ increases and thus quantum effects become more and more important in thermal transport. That is why the quantum phenomena are observable only at low temperatures. Note that the carbon materials have very high Debye temperature (diamond: 2230 K [3]). Therefore room temperature actually corresponds to low-temperature region for carbon systems.

3.1.2.2 The phonon mean free path l

Phonon mean free path is the average transport distance of phonon between two successive scatterings. l is determined by two types of processes: one is geometric scattering and the other is many-body scattering, Geometric scattering concerns phonon scatterings caused by structural imperfections, such as boundary scattering, interface scattering, defect scattering, impurity scattering, and so on. Additionally, the inhomogeneous distribution of atomic mass can also induce phonon scattering, such as isotope effects. In contrast to the weak temperature-dependence of geometric scattering, the strength of many-body scattering significantly varies as the temperature changes. The many-body scattering arises due to the coupling of phonons with other quasi-particles (e.g., electrons) and with collective excitations (e.g., photons), as well as with phonons themselves. Phonon–phonon scatterings are caused by anharmonic interactions. They can be categorized into normal scatterings (N-processes) and Umklapp (U) processes. Both are important for establishing local thermal equilibrium, but only U-processes contribute to thermal resistance. The many-body scatterings require the conservation of energy and quasi-momentum. The satisfaction of these conditions, however, becomes difficult in low dimensional systems like graphene or graphene nanoribbons (GNRs). Consequently, many-body scatterings become less important in systems of lower dimension.

According to kinetic theory of gases, there exists a relation between (average) phonon mean free path and thermal conductivity κ:

$$\kappa = cvl/D, \tag{3.3}$$

where c is the heat capacity per volume, v is the average group velocity of phonons and D is the dimensionality of the transport system. The relation is generally used for estimating (average) phonon mean free path.

When the system size becomes comparable or smaller than phonon mean free path, quantum effects becomes important. The graphene-like structures have very long phonon mean free path, as we will show. Theoretically studying such systems requires incorporating quantum effects into thermal transport. The non-equilibrium Green's function (NEGF) method, based on quantum mechanics, is the method of choice.

3.1.3 *Different Transport Regions*

Phonon mean free path is the key quantity for classifying transport regions. Thermal transport is diffusive for systems with sizes much larger than phonon mean free path. This is the classical case where Fourier's law is satisfied.

$$j = -\kappa \nabla_x T, \tag{3.4}$$

where j is the heat flux, κ is material's thermal conductivity, and $\nabla_x T$ is the temperature gradient along the transport direction (x direction).

When the system size is much smaller than phonon mean free path, phonons experience almost no scatterings during transport. In this case, thermal transport is quasi-ballistic or ballistic. Ballistic thermal transport exhibits a lot of unexpected, interesting features. For instance, there is no temperature gradient for ballistic thermal transport, thus Fourier's law fails. In such systems, thermal conductivity is not a constant but linearly dependent on the transport length. Instead, thermal conductance is a well-defined quantity and does not change as the system length increases.

The diffusive region, that is, the classical limit, has been well studied until now. Benefitting from the development of mesoscopic transport theory, ballistic transport can be well described, for instance, by the Landauer formalism to be shown later. However, much less is known for the transition region between ballistic and diffusive regions. There will be a transition from ballistic to diffusive transport when increasing system size. Increasing the temperature would also induce a ballistic-diffusive transition, since phonon mean free path is temperature-dependent. To understand how thermal-transport changes from the quantum limit to the classical limit, we need to investigate the transition region. The corresponding method development will be a promising direction for future research.

3.1.4 *The Landauer Formalism*

The Landauer formalism [4], first proposed to describe mesoscopic electronic transport, can be generally used for studying mesoscopic

Figure 3.1 Schematic diagram of a phonon transport system. Phonons (denoted by the circles) transport from the hot reservoir to the cold reservoir as well as in the inverse direction. This results in a net thermal current from the hot to the cold reservoir.

transport of quasi-particles and collective excitations. Here we show the Landauer formalism for phonon transport.

Figure 3.1 schematically shows a representative thermal-transport system in which the heat carriers are phonons. The dispersion relation $\omega(k)$ determines the intrinsic properties of phonons. One phonon of frequency ω and wave vector k has energy of $\hbar\omega$ and a group velocity of $\partial\omega/\partial k$ This phonon will experience scatterings during the transport process. Assuming that it has a transmission probability of $\Xi(\omega, k)$, the thermal current contributed by this phonon mode $I_{(\omega,k)}$ is

$$I_{(\omega,k)} = \hbar\omega\frac{\partial\omega}{\partial k}\Xi(\omega, k)[f(\omega, T_{\mathrm{H}}) - f(\omega, T_{\mathrm{C}})], \qquad (3.5)$$

where f is the Bose–Einstein distribution function, T_{H} and T_{C} represent the temperature of the hot and cold thermal reservoirs, respectively. The total thermal current I is the summation of the contribution of all the phonon modes:

$$I - \int \frac{dk}{2\pi} I_{(\omega,k)} = \int \frac{d\omega}{2\pi}\hbar\omega\Xi(\omega)[f(\omega, T_{\mathrm{H}}) - f(\omega, T_{\mathrm{C}})]. \qquad (3.6)$$

Here the transmission function $\Xi(\omega)$ is defined as

$$\Xi(\omega) = \sum_{k,\partial\omega/\partial k>0} \Xi(\omega, k). \qquad (3.7)$$

Note that only the phonon modes of positive group velocities are counted in the definition of $\Xi(\omega)$. The thermal conductance $\sigma(T)$ is given by the temperature derivative of I:

$$\sigma(T) = \frac{\hbar^2}{2\pi k_B T^2} \int_0^\infty d\omega \frac{\omega^2 e^{\hbar\omega/(k_B T)}}{(e^{\hbar\omega/(k_B T)} - 1)^2} \Xi(\omega), \tag{3.8}$$

where k_B is the Boltzmann constant.

Then we show that many important features of mesoscopic thermal transport can be revealed by analyzing the Landauer formula for thermal conductance (see Eq. 3.8). For ballistic transport, $\Xi(\omega)$ is equal to the number of phonon modes of frequency ω with positive group velocity. $\Xi(\omega)$ is an integer number and will not change when the transport length L is increasing. Thus, $\sigma(T)$ is constant (independent of L) at a given temperature. The thermal conductivity $\kappa \propto \sigma L$ is proportional to L and diverges for infinite large L.

Thermal conductance is a weighted integral of phonon transmission function, as shown by the Landauer formula. The weighting factor is strongly dependent on the phonon frequency. Defining $x = \hbar\omega/(k_B T)$, Eq. 3.8 changes into

$$\sigma(T) = \frac{k_B^2 T}{h} \int_0^\infty dx \frac{x^2 e^x}{(e^x - 1)^2} \Xi\left(\frac{k_B T}{\hbar} x\right). \tag{3.9}$$

Figure 3.2 presents the weighting factor $(x^2 e^x/(e^x - 1)^2)$ as a function of x. It clearly shows that the weighting factor deceases very fast with an increasing x. When $x > 10$, the weighting factor is nearly zero. This indicates that lower-frequency phonons have much larger contribution to thermal conductance than higher-frequency phonons. The phonons of very high frequencies, $\omega < 10 k_B T/\hbar$, have negligible contribution to thermal conductance. At very low temperatures (T close to zero), only those acoustic phonons of nearly zero ω are excited by the temperature. They can give quantized thermal conductance as we will discuss next.

3.1.5 *Quantized Thermal Conductance*

When the transmission function $\Xi(\omega) \equiv 1$, Eq. 3.9 gives the quantum of thermal conductance as

$$\sigma_0 = \frac{\pi^2 k_B^2 T}{3h}. \tag{3.10}$$

$\sigma_0 = (9.456 \times 10^{-13} \mathrm{W/K^2})T$ represents the maximum possible value of energy transported per phonon mode [5, 6]. Different from

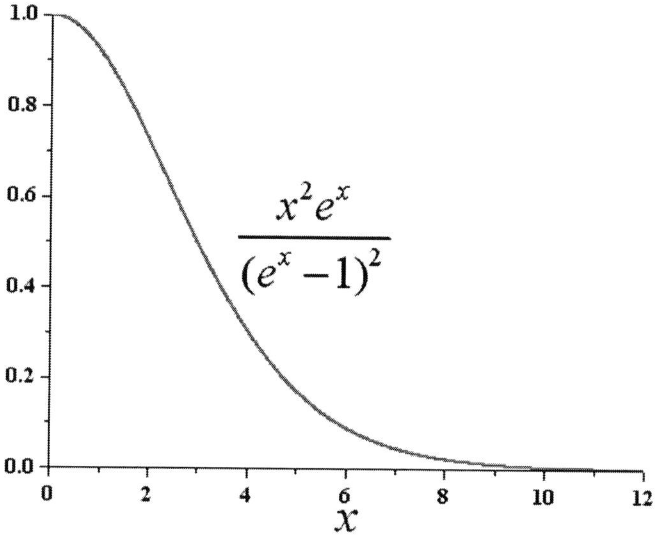

Figure 3.2 The weighting factor in the Landauer formula for thermal conductance (Eq. 3.9).

the quantum of electrical conductance $g_0 = e^2/h$, σ_0 is not a constant but linearly proportional to the temperature. A ballistic transport channel of electron contributes a quantum of electrical conductance accompanied by a quantum of thermal conductance. It is easy to prove that g_0 and σ_0 satisfy the Wiedemann–Franz Law:

$$\frac{\sigma_0}{g_0} = \frac{\pi^2 k_B^2 T}{3e^2} = L_0 T. \tag{3.11}$$

The quantum of thermal conductance is universal, independent of the characteristic of material. Moreover, it is also independent of particle statistics (i.e., universal for fermions, bosons, and anyons) [6]. The quantum of thermal conductance is first predicted theoretically [5]. However, its experimental evidence is extremely difficult, mainly because

1. Very low temperature is required for the observation of quantized thermal conductance. At normal temperatures thermal conductance will not be quantized, since a lot of phonon modes are simultaneously excited and all of them contribute to thermal conductance. It is necessary to decrease the temperature till

that all optical phonons are not excited. For that case, only those acoustic phonons with frequencies close to zero are excited and each of them contributes one quantum of thermal conductance. For quasi-one-dimensional (quasi-1D) systems (such as, nanowires, nanotubes, and nanoribbons), there are four eigenmodes of zero frequency, originating from three rigid translation invariances and one rigid rotation invariance. Consequently, their thermal conductance is $4\sigma_0$ at nearly zero temperature.

2. Accurately measuring σ_0, which is a very small quantity at low temperatures, requires extremely sensitive calorimeters and efficient approaches to exclude influence of environment, and thus is very challenging in experiments. As one of the most important breakthroughs in the research of quantum thermal transport, Schwab *et al.* measured thermal conductance of suspended insulating nanostructures at very low temperatures (below 1 K), and for the first time experimentally proved the existence of quantized thermal conductance [6].

3.2 The Non-equilibrium Green's Function Method

In this chapter, we will systematically discuss the NEGF formalism and try to offer a complete theory for the investigation of quantum thermal transport. The NEGF method is widely and successfully used for the study of electronic transport. Here the method is generalized for thermal transport. Since we focus on the thermal transport contributed by phonons, the NEGF represents the NEGF of phonons here.

3.2.1 *Hamiltonian of Thermal-Transport Systems*

Figure 3.3, using graphene junction as an example, shows a representative thermal-transport system. The center part (C) is connected to two thermal leads on the left (L) and right (R). The thermal leads are composed of semi-infinite ideal systems, in which phonon transport experiences no geometric scattering. The inclusion of many-body interactions in thermal leads on the one

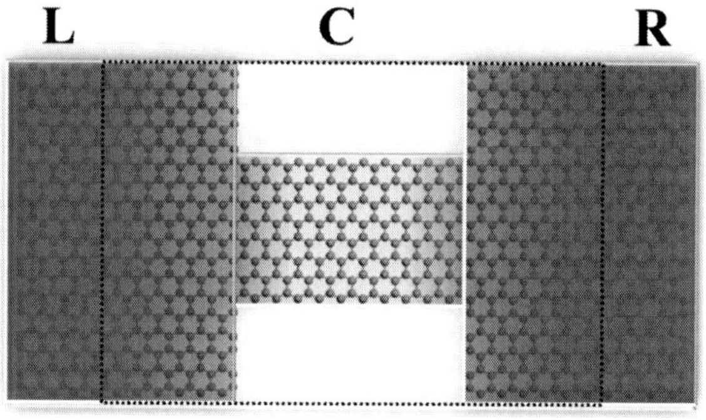

Figure 3.3 A representative thermal-transport system composed of a center part (C) as well as thermal leads on the left (L) and right (R).

hand is very difficult for the present theory, on the other hand is unnecessary because we are mainly interested in scatterings happened in the center part. Therefore, for simplicity we assume that all the phonon scatterings occur in the center part only.

Since phonons are the thermal-transport carriers here, the Hamiltonian (H) of thermal-transport systems is in fact the Hamiltonian of phonon systems.

$$H = \sum_{\alpha=\mathrm{L,C,R}} H_\alpha + V + H_n. \tag{3.12}$$

H_{L} H_{C}, and H_{R} are the harmonic Hamiltonian of the decoupled left, center and right parts of the transport system.

$$H_\alpha = \frac{1}{2}\left(\dot{u}_\alpha\right)^T \dot{u}_\alpha + \frac{1}{2} u_\alpha^T D_{\alpha\alpha} u_\alpha \quad (\alpha = \mathrm{L,\ C,\ R}), \tag{3.13}$$

where $u_\alpha = (u_1^\alpha, u_2^\alpha, u_3^\alpha, \cdots)^T$; $u_i^\alpha = \sqrt{M_i^\alpha} x_i^\alpha$; M_i^α and x_i^α respectively represent atomic mass and displacement of the ith degree of freedom in the α part; $\dot{u}_\alpha$ is the conjugated momentum of u_α; D is the mass-reduced force constant matrix.

$$D = \begin{pmatrix} D_{\mathrm{LL}} & D_{\mathrm{LC}} & 0 \\ D_{\mathrm{CL}} & D_{\mathrm{CC}} & D_{\mathrm{CR}} \\ 0 & D_{\mathrm{RC}} & D_{\mathrm{RR}} \end{pmatrix}. \tag{3.14}$$

Here, by default the coupling between the left and the right thermal leads are zero. In principle, this condition can be satisfied always by enlarging the size of the center part so as to decouple the two thermal leads from each other. The mass-reduced force constant is the mass-weighted second derivative of energy with respect to displacement at the equilibrium positions.

$$[D_{\alpha\beta}]_{ij} = D_{ij}^{\alpha\beta} = \frac{\partial^2 E}{\partial u_i^\alpha \partial u_j^\beta}|_0. \tag{3.15}$$

The term V describes the harmonic coupling between the center part with the two thermal leads.

$$V = (u_{\mathrm{L}})^T\ D_{\mathrm{LC}}u_{\mathrm{C}} + (u_{\mathrm{C}})^T\ D_{\mathrm{CR}}u_{\mathrm{R}}. \tag{3.16}$$

The term H_n represents the nonlinear interactions or the many-body interactions. Here it denotes the anharmonic interactions among the center part. It contains the high order (higher than 2) derivatives of energy with respective to displacement at the equilibrium positions.

$$H_n = \sum_{n=3,4,\cdots} H^{(n)} = \sum_{n=3,4,\cdots} \frac{1}{n!} \sum_{i_1,i_2,\cdots,i_n} D_{i_1,i_2,\cdots,i_n} u_{i_1} u_{i_2} \cdots u_{i_n} \tag{3.17}$$

where

$$D_{i_1,i_2,\cdots,i_n} = \frac{\partial^n E}{\partial u_{i_1} \partial u_{i_2} \cdots \partial u_{i_n}}|_0. \tag{3.18}$$

If the interactions are harmonic, there will be no phonon–phonon interactions. The presentence of H_n will cause phonon–phonon scatterings. The nth order term $H^{(n)}$ corresponds to interactions involving n phonons. Most theoretical treatments, if considering anharmonic interactions, will include H_n till the 4th term. This is because the higher order terms usually have negligible effects on thermal transport and dealing with those terms is extremely troublesome. While it is important to including the 4th term that helps stabilizing the system. The inclusion of the 3rd anharmonic term only sometimes causes the system energetically unstable (i.e., absence of lower boundary for the energy).

3.2.2 *The NEGF Formalism*

The NEGF formalism is a powerful tool for dealing with many-body interactions. Two types of formalisms are widely used: the contour-ordered Green's function and the real-time Green's function. They are equivalent to each other. A detailed discussion of the NEGF formalisms is beyond the scope of this book. The rigorous description of the formalisms can be found in various literatures [7–9]. Here, we will briefly introduce the real-time Green's function and present related formula for thermal transport.

3.2.2.1 Six real-time Green's functions

There are six real-time Green's functions. They are defined as:

1. The time-ordered Green's function $G_{ij}^{t} = -\frac{i}{\hbar}\left\langle T_t u_i\left(t\right) u_j\left(t'\right)\right\rangle$
2. The anti-time-ordered Green's function $G_{ij}^{\bar{t}} = -\frac{i}{\hbar}\left\langle T_{\bar{t}} u_i\left(t\right) u_j\left(t'\right)\right\rangle$
3. The "greater than" Green's function $G_{ij}^{>} = -\frac{i}{\hbar}\left\langle u_i\left(t\right) u_j\left(t'\right)\right\rangle$
4. The "less than" Green's function $G_{ij}^{<} = -\frac{i}{\hbar}\left\langle u_j\left(t'\right) u_i\left(t\right)\right\rangle$
5. The retarded Green's function $G_{ij}^{r} = -\frac{i}{\hbar}\theta\left(t-t'\right)\left\langle\left[u_i\left(t\right), u_j\left(t'\right)\right]\right\rangle$
6. The advanced Green's function $G_{ij}^{a} = +\frac{i}{\hbar}\theta\left(t'-t\right)\left\langle\left[u_i\left(t\right), u_j\left(t'\right)\right]\right\rangle$

Here $i, j = 1, 2, 3\ldots, 3N$ are atomic indexes (N is number of atoms in the system); t and t' represent the real time; T_t and $T_{\bar{t}}$ respectively denote the time-ordered and the anti-time-ordered operators; u is the mass-weighted displacement as defined previously; θ is the unit step function; the square brackets are the commutators; $\langle\ldots\rangle$ represents assembling average over the density matrix.

Based on the definition, there are some relations between the six real-time Green's functions.

$$G^{t} + G^{\bar{t}} = G^{>} + G^{<} \tag{3.19}$$

$$G^{r} = G^{t} - G^{<} \tag{3.20}$$

$$G^{a} = G^{<} - G^{\bar{t}} \tag{3.21}$$

Among the six Green's functions, only three of them are linearly independent. Selecting which kind of Green's functions for suitable for practical use is a little bit arbitrary. However, the mostly used Green's functions are:

1. $G^>$ and $G^<$: they are usually regarded as correlation functions, and are directly related to observable physical quantities (e.g., thermal current, density of states and so on) and dynamic properties.
2. G^r and G^a: they have good analytical properties, facilitate calculations of physical response, and also have simple relations to observable physical quantities.

In equilibrium or steady states, the system is time-translation invariant. The Green's functions depend only on the difference in time. Then it is more convenient to go from time space to frequency space through Fourier transformation:

$$G[\omega] = \int_{-\infty}^{+\infty} G(t)\, e^{i\omega t} dt. \tag{3.22}$$

In frequency space, there are relations:

$$G^<[-\omega] = G^>[\omega]^T \tag{3.23}$$

$$G^<[\omega]^\dagger = -G^<[\omega] \tag{3.24}$$

$$G^r[-\omega] = G^r[\omega]^* \tag{3.25}$$

$$G^r[\omega]^\dagger = G^a[\omega] \tag{3.26}$$

For steady states, only two Green's functions are linearly independent. This number decreases to one for equilibrium states, since equilibrium systems satisfy the fluctuation–dissipation theorem, which gives rise to an additional relation between Green's functions.

$$G^<[\omega] = f(\omega)\left(G^r[\omega] - G^a[\omega]\right), \tag{3.27}$$

where f is the Bose–Einstein distribution function. The correlation function $G^<$ contains information of fluctuation, while $G^r - G^a$ describes dissipation of the system. The fluctuation–dissipation theorem tells us that fluctuation is proportional to dissipation in equilibrium states.

3.2.2.2 The Dyson equation

The Dyson equation is expressed as

$$G = G_0 + G_0 \Sigma G, \tag{3.28}$$

where G and G_0, respectively, are the Green's function of interacting and noninteracting systems, which have Hamiltonians of $H = H_0 + H'$ and H_0, respectively. The self energy Σ includes all the effects of interactions (H'). In the following, we will focus on steady or equilibrium states and prefer to express quantities in the frequency space.

Employing the Langreth theorem [7], the Dyson equations of G^r and $G^<$ in frequency spaces can be written as

$$G^r[\omega] = G_0^r[\omega] + G_0^r[\omega]\Sigma^r[\omega]G^r[\omega] \tag{3.29}$$

$$G^<[\omega] = G^r[\omega]\left(G_0^r[\omega]\right)^{-1} G_0^<[\omega]\left(G_0^a[\omega]\right)^{-1} G^a[\omega] + G^r[\omega]\Sigma^<[\omega]G^a[\omega] \tag{3.30}$$

They are the two basic equations for the NEGF formalism of thermal transport.

3.2.2.3 Basic equations of NEGF

In an open thermal-transport system as shown in Fig. 3.3, the degree of freedom is infinite for the whole system. Directly dealing with this infinitely large system is impossible for practical calculations. Since we are only interested in the center part where phonon scatterings occur, we can simplify the transport problem to finite degree of freedom and focus on calculating the Green's function of the center part G_{CC}. As we will show later, this can be achieved by projecting the influence of thermal leads into the center part. Once we get G_{CC}, thermal-transport properties of the whole system can be obtained. If there is no likelihood of confusion, hereafter we drop the subscript "CC" and use G to represent G_{CC} for simplicity.

In the following, we discuss basic equations of NEGF for two different cases: *without* and *with* many-body interactions (i.e., anharmonic interactions here).

Case (1): $H_n = 0$

Define G_0 as the Green's function of the center part *without* anharmonic interactions. When the center part is decoupled with thermal leads, its Green's function g satisfies the equation

$$g^r[\omega] = \left[(\omega + i\delta)^2 I - D_{\rm CC}\right]^{-1}. \tag{3.31}$$

The influence of thermal leads is projected into the center part in the form of self energy Σ. Each thermal lead contributes a self energy term

$$\Sigma = \Sigma_{\rm L} + \Sigma_{\rm R} \tag{3.32}$$

The self energy of thermal leads is written as

$$\Sigma_\alpha = D_{{\rm C}\alpha} g_{\alpha\alpha} D_{\alpha{\rm C}} \quad (\alpha = {\rm L, R}), \tag{3.33}$$

where the surface Green's function $g_{\alpha\alpha}$ is the Green's function of isolated semi-infinite thermal lead. It satisfies

$$g^r_{\alpha\alpha}[\omega] = \left[(\omega + i\delta)^2 I - D_{\alpha\alpha}\right]^{-1} \quad (\alpha = {\rm L, R}). \tag{3.34}$$

The surface Green's function can be calculated by a simple iteration method [10] or by a more efficient decimation approach [11].

Applying the Dyson equation, G_0 can be calculated from g:

$$\begin{aligned} G^r_0[\omega] &= \left[(g^r[\omega])^{-1} - \Sigma^r[\omega]\right]^{-1} \\ &= \left[(\omega + i\delta)^2 I - D_{\rm CC} - \Sigma^r[\omega]\right]^{-1} \end{aligned} \tag{3.35}$$

$$\begin{aligned} G^<_0[\omega] &= G^r_0[\omega]\,(g^r[\omega])^{-1} g^<[\omega]\,(g^a[\omega])^{-1} G^a_0[\omega] \\ &\quad + G^r_0[\omega]\Sigma^<[\omega]G^a_0[\omega] \end{aligned} \tag{3.36}$$

For equilibrium states, fluctuation–dissipation theorem is satisfied and Eq. 3.36 can be further simplified into

$$G^<_0[\omega] = G^r_0[\omega]\Sigma^<[\omega]G^a_0[\omega]. \tag{3.37}$$

Case (2): $H_n \neq 0$

Define G as the Green's function of the center part *with* anharmonic interactions. The anharmonic interaction H_n gives rise to a many-body self energy term Σ_n. G can be obtained from G using the Dyson equation.

$$\begin{aligned} G^r[\omega] &= \left[\left(G^r_0[\omega]\right)^{-1} - \Sigma^r_n[\omega]\right]^{-1} \\ &= \left[(\omega + i\delta)^2 I - D_{\rm CC} - \Sigma^r[\omega] - \Sigma^r_n[\omega]\right]^{-1} \end{aligned} \tag{3.38}$$

It is easy to prove that

$$\left(G_0^r[\omega]\right)^{-1} G_0^<[\omega] \left(G_0^a[\omega]\right)^{-1} = \Sigma^<[\omega]. \tag{3.39}$$

Thus

$$G^<[\omega] = G^r[\omega]\left(\Sigma^<[\omega] + \Sigma_n^<[\omega]\right) G^a[\omega]. \tag{3.40}$$

As a short summary, we list basic equations of NEGF:

$$G_0^r[\omega] = \left[(\omega + i\delta)^2 I - D_{CC} - \Sigma^r[\omega]\right]^{-1}$$

$$G_0^<[\omega] = G_0^r[\omega]\Sigma^<[\omega]G_0^a[\omega]$$

$$G^r[\omega] = \left[(\omega + i\delta)^2 I - D_{CC} - \Sigma^r[\omega] - \Sigma_n^r[\omega]\right]^{-1}$$

$$G^<[\omega] = G^r[\omega]\left(\Sigma^<[\omega] + \Sigma_n^<[\omega]\right) G^a[\omega]$$

The self energy of thermal leads Σ is obtained by calculating the surface Green's function. While the many-body self energy Σ_n contributed by anharmonic interactions can be calculated by many-body perturbation approaches, for instance by using Feynman diagram techniques [12–14].

3.2.2.4 Work flow of NEGF

Having introduced the NEGF formalism, we show the work flow of NEGF in practical calculations, as presented in Fig. 3.4. Here we assume that the Hamiltonian is known. How to obtain the Hamiltonian will be discussed later. The first step is to calculate the surface Green's function and the self energy of thermal leads. Then applying Eq. 3.35 and Eq. 3.37, Green's functions without phonon–phonon interactions (G_0^r and $G_0^<$) are obtained.

If anharmonic interactions are included in the calculation, further calculations are required to obtain full Green's functions (G^r and $G^<$). There are two options for such calculations: non-self-consistent calculation and self-consistent iteration. Both options start from G_0^r and $G_0^<$, and use them to construct many-body self energies (Σ_n^r and $\Sigma_n^<$) by applying the Feynman diagram techniques, then calculate full Green's functions according to Eqs. 3.38 and 3.40. The non-self-consistent calculation terminates at this step. Differently, the self-consistent iteration approach uses the newly

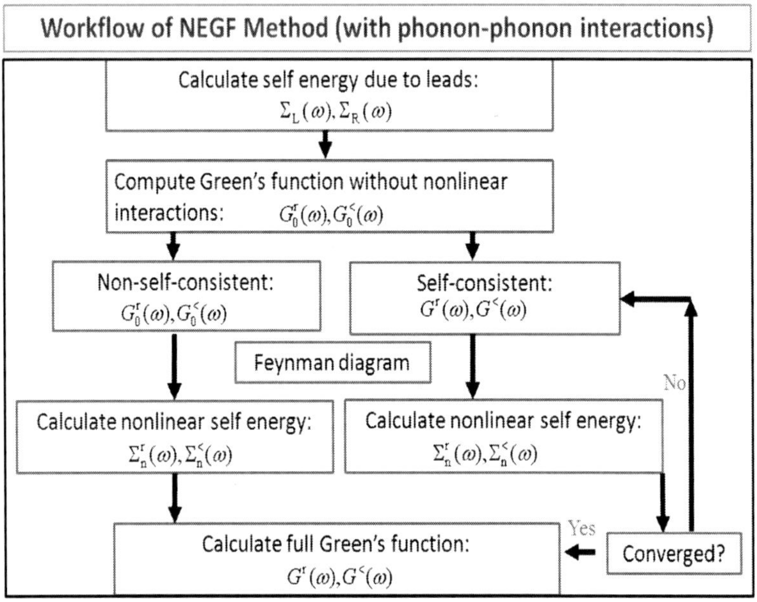

Figure 3.4 Work flow of the NEGF method for the case of phonon–phonon interactions are included in the calculation.

obtained Green's functions to construct many-body self-energies again, and iterations are done till getting converged G^r and $G^<$.

Usually we neglect the term Σ_n in NEGF calculations, because computing Σ_n is complex and only feasible for small model systems. This is a safe approximation for general uses because the influence of anharmonic interactions is only important at high temperatures.

3.2.3 *NEGF and Thermal-Transport Properties*

Once getting the Green's functions, we can obtain all the one-particle quantities relevant to thermal transport, including phonon density of states (DOS), thermal current, thermal conductance, phonon transmission function and so on. Consequently, a complete physical picture of quantum thermal transport can be established. In the following, we show the relation between NEGF and thermal-transport properties.

3.2.3.1 Phonon DOS

Assume that the system has $3N$ degree of freedom. The nth phonon mode has a frequency of ω_n and an eigenvector of $|u^n\rangle$ whose i^{th} component is u_i^n. Then

$$D\left|u^n\right\rangle = \omega_n^2\left|u^n\right\rangle. \tag{3.41}$$

Define phonon DOS as

$$\rho(\omega) = \sum_n \delta(\omega - \omega_n) \tag{3.42}$$

and phonon local DOS (LDOS) of the ith degree of freedom as

$$\rho_i(\omega) = \sum_n \left|u_i^n\right|^2 \delta(\omega - \omega_n). \tag{3.43}$$

The total phonon DOS is the summation of phonon LDOS over all the degree of freedom

$$\rho(\omega) = \sum_i \rho_i(\omega). \tag{3.44}$$

The total phonon DOS satisfies the relation

$$\int_0^\infty d\omega\rho(\omega) = 3N \tag{3.45}$$

On the other hand, the phonon DOS can also be given by NEGF. We know that

$$\begin{aligned} G^r[\omega] &= \frac{1}{(\omega + i\delta)^2 - D} \\ &= \sum_n \frac{|u^n\rangle\langle u^n|}{(\omega + i\delta)^2 - \omega_n^2} \\ &= \sum_n \frac{|u^n\rangle\langle u^n|}{2(\omega + i\delta)}\left(\frac{1}{\omega + i\delta - \omega_n} + \frac{1}{\omega + i\delta + \omega_n}\right). \end{aligned} \tag{3.46}$$

The formula can be further simplified, thus

$$-\frac{2\omega}{\pi}(\text{Im}G^r[\omega])_{ii} = \sum_n \left|u_i^n\right|^2 [\delta(\omega - \omega_n) + \delta(\omega + \omega_n)]. \tag{3.47}$$

When $\omega > 0$, $\delta(\omega + \omega_n) = 0$. The phonon LDOS can be expressed as

$$\rho_i(\omega) = -\frac{2\omega}{\pi}(\text{Im}G^r[\omega])_{ii}, \tag{3.48}$$

and the total phonon DOS is

$$\rho(\omega) = -\frac{2\omega}{\pi}\mathrm{Tr}(\mathrm{Im}G^r[\omega]). \tag{3.49}$$

Although Eqs. 3.48 and 3.49 are obtained from a special case of no phonon–phonon interactions, the same formulae work also for cases including phonon–phonon interactions providing that the quasi-particle picture does not break down. The phonon DOS and LDOS gives the distribution of phonons in frequency space and in real space. This kind of information is very helpful to the analysis of quantum thermal-transport processes.

3.2.3.2 Thermal current

First, we would like to show two useful functions in the NEGF formulism:

(1) the spectral function

$$A[\omega] = i\left(G^r[\omega] - G^a[\omega]\right) = i\left(G^>[\omega] - G^<[\omega]\right). \tag{3.50}$$

(2) the Γ function describing scattering rate of phonons

$$\begin{aligned}\Gamma[\omega] &= i\left(\Sigma^r[\omega] - \Sigma^a[\omega]\right) = \Gamma_L[\omega] + \Gamma_R[\omega]\\ \Gamma_n[\omega] &= i\left(\Sigma^r_n[\omega] - \Sigma^a_n[\omega]\right)\end{aligned} \tag{3.51}$$

There is an important relation between them:

$$A[\omega] = G^r[\omega]\left(\Gamma[\omega] + \Gamma_n[\omega]\right)G^a[\omega]. \tag{3.52}$$

For the case of no phonon–phonon interaction, this relation changes into

$$A_0[\omega] = G^r_0[\omega]\,\Gamma[\omega]\,G^a_0[\omega] \tag{3.53}$$

The thermal current (for the left thermal lead) can be expressed as

$$I_{\mathrm{L}} = -\int_{-\infty}^{+\infty}\frac{d\omega}{2\pi}\hbar\omega\mathrm{Tr}\left(G^r[\omega]\,\Sigma^<_{\mathrm{L}}[\omega] + G^<[\omega]\,\Sigma^a_{\mathrm{L}}[\omega]\right). \tag{3.54}$$

The formula, after symmetrization, can be written as

$$I_L = \int_0^{+\infty} d\omega i_{\mathrm{L}}[\omega], \tag{3.55}$$

where $i_L[\omega]$ is the thermal current spectral function of the left thermal lead:

$$i_{\mathrm{L}}[\omega] = \frac{i\hbar\omega}{2\pi}\mathrm{Tr}\left(A[\omega]\Sigma_{\mathrm{L}}^{<}[\omega] - G^{<}[\omega]\Gamma_{\mathrm{L}}[\omega]\right). \tag{3.56}$$

Similarly, the thermal current spectral function of the right thermal lead is

$$i_{\mathrm{R}}[\omega] = \frac{i\hbar\omega}{2\pi}\mathrm{Tr}\left(A[\omega]\Sigma_{\mathrm{R}}^{<}[\omega] - G^{<}[\omega]\Gamma_{\mathrm{R}}[\omega]\right). \tag{3.57}$$

The two thermal leads always give same total current for steady states. However, their spectral functions can be different if asymmetry exists. A thorough investigation of thermal rectification requires such information.

3.2.4 *Thermal Conductance and Phonon Transmission*

Thermal conductance is defined as

$$\sigma = \lim_{\Delta T\to 0}\frac{I}{\Delta T}, \tag{3.58}$$

where ΔT is the temperature difference between the two thermal leads, assuming that the temperature of the left and right thermal leads are $T + \Delta T/2$ and $T - \Delta T/2$, respectively. We introduce variational derivative for Green's function as

$$\begin{aligned}\frac{\delta G}{\delta T} &= \lim_{\Delta T\to 0}\frac{F(T_{\mathrm{L}}, T_{\mathrm{R}}) - F(T, T)}{T_{\mathrm{L}} - T_{\mathrm{R}}}\\ &= \lim_{\Delta T\to 0}\frac{F(T+\Delta T/2, T-\Delta T/2) - F(T, T)}{\Delta T}\end{aligned} \tag{3.59}$$

Similar variational derivative is introduced for other functions, like self energy and thermal current. As variational derivative of thermal current with respect to temperature, thermal conductance can be expressed as

$$\sigma = \int_0^{\infty}\frac{d\omega}{2\pi}\hbar\omega\frac{\partial f}{\partial T}\tilde{\Xi}(\omega), \tag{3.60}$$

where $\tilde{\Xi}(\omega)$ is the effective phonon transmission function.

The effective phonon transmission of the left thermal lead is

$$\tilde{\Xi}_{\mathrm{L}}(\omega) = \mathrm{Tr}\left[G^{r}(\Gamma_{\mathrm{R}} + \frac{1}{2}\Gamma_{n} + S)G^{a}\Gamma_{\mathrm{L}}\right], \tag{3.61}$$

where

$$S = \left(f \frac{\delta \Gamma_n}{\delta T} - i \frac{\delta \Sigma_n^<}{\delta T} \right) \left(\frac{\partial f}{\partial T} \right)^{-1}. \tag{3.62}$$

Similarly, the effective phonon transmission of the right thermal lead is

$$\tilde{\Xi}_{\mathrm{R}}(\omega) = \mathrm{Tr}\left[G^r (\Gamma_{\mathrm{L}} + \frac{1}{2}\Gamma_n - S) G^a \Gamma_{\mathrm{R}} \right]. \tag{3.63}$$

The effective phonon transmission could be different due to the existence of asymmetry. However, they give the same thermal conductance.

3.2.5 *The NEGF Method and the Landauer Formalism*

Superficially, the NEGF method looks quite different from the Landauer formalism. The NEGF method focuses on the scattering region (i.e., the center part of the transport system), calculates its Green's functions, and then uses them to describe thermal-transport properties of the whole transport system. In contrast, the Landauer formalism does not care about the details of the center part. Instead, the key information for the Landauer formalism is the phonon transmission function that is the probability of a phonon modes transporting from one thermal lead to the other one.

The advantage of the Landauer formalism is rooted in its simplicity. It is easy to use. At the same time, it provides a clear physical picture for understanding mesoscopic transport problems. However, the Landauer formalism itself cannot provide the phonon transmission function, thus its application requires input from other approaches. As a comparison, the NEGF method is conceptually more complex. The NEGF method can deduce the Landauer formalism as well as the phonon transmission. In this sense, the NEGF method is more fundamental than the Landauer formalism.

Then we show the relation between the NEGF method and the Landauer formulism. Thermal conductance derived by the NEGF method is presented in Eq. 3.60. This is actually of the same form as the Landauer formalism. Meanwhile, the expression of phonon transmission function is provided by Eqs. 3.60 and 3.63.

For cases without phonon–phonon interactions, phonon transmission functions of the left and right thermal leads are, respectively,

$$\Xi_{\mathrm{L}}(\omega) = \mathrm{Tr}\,[G^r\Gamma_{\mathrm{R}}G^a\Gamma_{\mathrm{L}}] \tag{3.64}$$

and

$$\Xi_{\mathrm{R}}(\omega) = \mathrm{Tr}\,[G^r\Gamma_{\mathrm{L}}G^a\Gamma_{\mathrm{R}}]\,. \tag{3.65}$$

For cases with phonon–phonon interactions, we should adopt the effective phonon transmission function shown in Eqs. 3.61 and 3.63. The two transmission functions could be different, but they give the same total thermal current. For thermal rectification effect, it is meaningful and important to investigate the differences between the two transmission functions.

3.2.6 *First-Principles-Based NEGF*

We have systematically introduced the NEGF method, from basic theory to practical calculation. Starting from the Hamiltonian, the NEGF method can give all the one-particle observables of quantum thermal transport. However, the NEGF method itself is not first-principles, because the Hamiltonian, the most important quantity of the method, requires external input. Following we discuss the calculation of the Hamiltonian of thermal-transport systems.

The Hamiltonian of phonons is actually composed of force constants, that is, the derivatives of energy with respective to displacement at equilibrium positions. Computing the Hamiltonian by using empirical potential is very efficient and feasible for large systems. Successful empirical potentials are now available for few elements, such as C, O, H, and Si. However, the accuracy of the results strongly relies on the employed empirical potential parameters.

First-principles approaches, based on quantum mechanics, can predict various materials' properties, without the need of any empirical parameters. The first-principles approach we mention is based on the Kohn–Sham density functional theory (DFT). DFT has been widely and successfully used in the fields of physics, chemistry, material science, biology and so on. It turns out to be an important research tool for predicting materials' electronic, mechanical, thermal, magnetic, and optical properties.

DFT offers a promising way to investigate the ground state of many-body systems. In DFT, the unsolvable many-body problem is simplified into a solvable one-particle problem under effective external potential. All the complexity arising from many-body interactions is contained in the exchange-correlation (XC) functional. The exact form of XC functional is unknown and needs to be approximated. The success of DFT relies on the fact that even very simple approximations (local density approximation and generalize gradient approximation) to XC functional work well for most of systems.

Using first-principles approaches, the Hamiltonian of phonons can be calculated for realistic materials with high accuracy. Based on the obtained Hamiltonian, the NEGF method can provide a complete description on quantum thermal transport. Therefore, by combining first-principles approaches with the NEGF method, a first-principles prediction or computational experiment on thermal-transport properties of realistic materials becomes feasible. The first-principles-based NEGF is a powerful tool in the research of quantum thermal transport.

3.3 Thermal Conduction of Graphene: Experiment

Graphene, a monolayer of sp^2-bonded carbon atoms densely packed in a honeycomb crystal structure (Fig. 3.5), is a conceptually new material, because (i) it is the first strictly 2D material that opens new

Figure 3.5 Graphene is a 2D material made of carbon atoms.

opportunities for the developments of low-dimensional physics, and (ii) the existence of massless Dirac Fermions in graphene, whose behavior is governed by Dirac's (relativistic) equation, led to the emergence of a new paradigm of "relativistic" condensed-matter physics [15]. In 2004, for the first time graphene was experimentally observed [16, 17] by Andre Geim and Konstantin Novoselov at the University of Manchester; they were awarded the Nobel Prize in Physics 2010 for their groundbreaking experiments regarding graphene. Since then an explosion of interest has been devoted to graphene. This unique material has been revealed to exhibit excellent electronic, thermal, magnetic, optical, and mechanical properties [15, 18]. In this chapter, we will briefly show experimental progresses on graphene, focusing on the thermal conduction related issues, including material growth, thermal measurement techniques, and main experimental results. Theoretical studies on thermal conduction of graphene will be discussed in the next chapter.

The first successful experimental production of graphene was achieved in 2004 by micro-mechanical alleviation of graphite [16], which is often referred as scotch tape or drawing method. The technique provides high mobility graphene with low defect concentration and thus was widely used in early-stage graphene experiments. However, the size of produced graphene is too small and the price is too high. Numerous approaches have been developed for massive production of high-quality graphene. Macroscopic-scale graphene sheets can be prepared by chemical approaches, such as from the exfoliation–reintercalation–expansion of graphite [19] or from the reduced graphene oxide [20]. The chemical derived graphene, if with improved quality, will find wide applications in future electronics. Other popular techniques include chemical vapour deposition (CVD) of hydrocarbons on metals (e.g., copper foils [21] and thin nickel layers [22]) and expitaxy/thermal decomposition of silicon carbide [23]. These progresses on growth techniques prompt the development of graphene-based applications.

Nanoscale thermal measurement is very challenging in experiments, since reliable experiments require high-accuracy control of material growth and manipulation as well as high-resolution

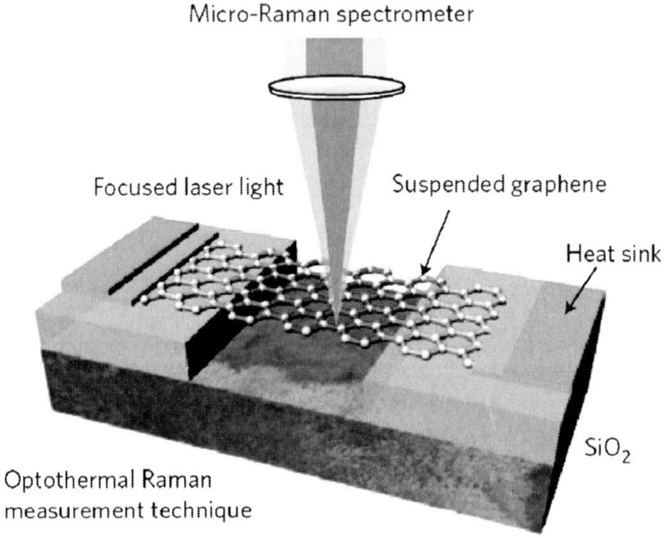

Figure 3.6 Schematic of the experimental setup for measuring the thermal conductivity of graphene using the optothermal Raman technique. Reprinted by permission from Macmillan Publishers Ltd: *Nature Materials* (Ref. [28]), copyright 2011.

characterization of spatial temperature profile. Many approaches have been developed for measuring thermal conduction of quasi-1D systems, such as nanotubes and nanowires. These approaches can be roughly classified into two groups: contact and non-contact approaches. In the former approaches, the measured samples are heated by electricity. The sample temperature is characterized by thermocouple [1], thermometer employing the resistance-temperature relation [24], or I–V curves of Joule self-heating experiments [25]. In the non-contact approaches, experimental samples are heated by electricity [26] or by laser [27]. The temperature profile is determined from the temperature-induced shifts in the G band Raman frequency, which is referred as optothermal Raman measurement technique. Most experiments measure thermal conductivity of suspended graphene using the optothermal Raman technique (depicted in Fig. 3.6) [28–33], and some experiments investigate thermal conductivity of supported graphene employing the contacted electrical approach [34].

The first experimental study measured thermal conductivity of suspended graphene exfoliated from highly oriented pyrolytic graphite (HOPG) using the optothermal Raman technique [29, 30]. There, extremely high thermal conductivity in the range from ~3000 W/m/K to ~5000 W/m/K at room temperature is reported. Moreover, phonon mean free path in graphene is estimated to be around 775 nm at room temperature [30]. Another experiment, using similar Raman technique but with an additional power meter under suspended graphene, reported thermal conductivity of suspended graphene grown by CVD exceeding (2500 + 1100/–1050) W/m/K near 350 K. The other experiment, using a combination of Raman spectroscopy and electrical current heating, found thermal conductivity of suspended graphene grown by CVD in the range from ~1500 W/m/K to ~5000 W/m/K near 350 K [33]. All these experiments consistently show that thermal conductivity of suspended graphene is extremely high, exceeding that of diamond which possesses the highest bulk thermal conductivity. Contacting graphene with a substrate significantly affects thermal-transport properties of graphene. A recent experiment found that the room temperature thermal conductivity of exfoliated graphene on a silicon dioxide support is only around 600 W/m/K, which is still higher than those of metals such as cooper [34]. This value is significantly lower than that of suspended graphene. The support-induced thermal conductivity reduction was attributed to strong interface scattering of flexural modes and phonon leaking across the interface [34]. A comprehensive review on thermal-transport properties of graphene-related materials can be found in Ref. [28].

3.4 Thermal Conduction of Graphene: Theory

In this chapter, we will present recent theoretical studies on thermal conduction of graphene. A complete review on the progress of the whole research field is obviously beyond the scope of the present chapter. Instead, we place extra emphasis on fundamental aspects of thermal transport in graphene, which could help readers understand the more complex and exciting cutting-edge research and development of the field. On the other hand, we have already shown the basic theory and numerical implementation of the NEGF method. Some representative applications of the NEGF method will be shown to demonstrate the powerfulness of this theoretical tool in exploring quantum thermal-transport phenomena. Graphene-based transport systems are very suitable for such purpose because of their simplicity and rich experimental data. The NEGF study will focus on thermal transport in quasi-1D systems whose size perpendicular to the transport direction is finite. This, to some extent, differs from most experiments whose samples have dimensions of macroscopic scale. However, results of wide enough systems can be used to compare with those experiments. An additional advantage of the quasi-1D study is that by systematically enlarging the system size, size effects and quantum confinement effects, which for experiments are challenging to explore, can be investigated.

At first we will systematically investigate thermal transport in GNRs and discuss the size effects and influence of edge shape on thermal-transport properties. Then we will discuss the origin of extremely high thermal conductivity in graphene. The NEGF calculations show that it is the very long phonon mean free path rather than extraordinary high ballistic thermal conductance that gives rise to superiorly high thermal conductivity in graphene. Finally we will systematically study thermal-transport properties of graphene-based nanodevices. The obtained structure-property relations offer guidelines to the design of graphene-based electronic, thermal, and thermoelectric nanodevices.

3.4.1 *Graphene Nanoribbons*

As the first application example for the NEGF method, we investigate thermal-transport properties of pristine GNRs. GNRs are the building blocks of graphene-based nanoelectronics. The experimental production of high-quality GNRs with controlled shape and width is now feasible, for instance, by graphite nanotomy process, which uses sharp diamond knife to cut graphite into graphite nanoblocks that are then exfoliated into GNRs [35], or by longitudinal unzipping of CNTs [36, 37]. Compared to extensive study on electronic transport, much less attention has been paid for understanding thermal transport properties of GNRs. Such study, however, is of crucial important for the development of future nanoelectronics because of the serious heat dissipation in ever-smaller integrated circuits.

It is well known that electronic properties of GNRs are strongly dependent on their edge shape and ribbon width. Such kind of knowledge is essential for applications of GNRs. Similarly, the corresponding information on thermal-transport properties is significant not only for improving performance of GNR-based electronic devices but also for designing GNR-based thermal management or thermoelectric devices. Therefore, we investigate thermal-transport properties of GNRs with different widths and edge shapes, using the NEGF method. Xu *et al.* [38] provide more details of this study.

It should be stressed that a quantum description of thermal transport is necessary for systems of short transport lengths (compared to phonon mean free path) or equivalently at low temperatures (compared to the Debye temperature). In the diffusive region (i.e., the length of the transport system is much larger than phonon mean free path), the Newton's mechanics-based approaches [e.g., the classical molecular dynamics (MD)] and the quantum mechanics-based approaches (e.g., the NEGF method) are equivalent and both work at this classical limit. While as the system size becomes comparable or smaller than phonon mean free path, the Newton's mechanics-based approaches fail because of the dominant quantum effects due to the wave nature of thermal-transport carriers, which cannot be correctly described in the classical framework. Alternatively, as the temperature decreases phonon mean free path increases thus the quantum effects play a more and more important role in thermal transport. For carbon materials, the room temperature belongs to the low-temperature region because of their extremely high Debye temperature (over 2000 K for diamond). A full quantum mechanics description is requisite for thermal transport in graphene around room temperature.

GNRs with width varying from 0.5 nm to 35 nm are considered for exploring size effects. The influence of edge shape will also be investigated by comparing thermal-transport properties of two representative types of GNRs: zigzag GNRs (ZGNRs) and armchair GNRs (AGNRs) as shown in Fig. 3.7. Following the conventional notation, N_a - AGNR(N_z - ZGNR) denotes an AGNR (a ZGNR) with

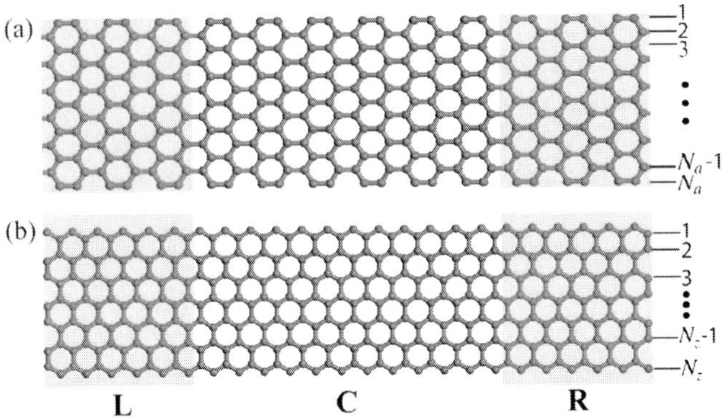

Figure 3.7 Schematic illustration of two-probe transport system, where the center transport region (C) is connected to left (L) and right (R) semi-infinite thermal leads. (a) N_a-AGNR, (b) N_z–ZGNR. Reprinted with permission from Ref. [38]. Copyright 2009, AIP Publishing LLC.

$N_a(N_z)$ carbon dimer lines (zigzag carbon chains) across the ribbon width. Since GNRs with larger ribbon width (W) support more phonon transport channels and thus gives higher thermal conductance (σ), we define scaled thermal conductance as thermal conductance per unit area (σ/S) for comparing thermal-transport ability among different GNRs or comparing with other materials. The cross sectional area S is defined to be $S = W\delta$, where $\delta = 0.335$ nm is chosen as the layer separation in graphite. Note that CNTs can be viewed as rolled-up GNRs. For the sake of comparison, we also consider armchair carbon nanotubes (ACNTs) and zigzag carbon nanotubes (ZCNTs), whose widths refer to their circumferences.

In the transport calculations, we first perform geometric optimization and then calculate force constants using the second-generation reactive empirical bond order potential [39]. This empirical potential is widely used for simulating materials composed of carbon and hydrogen. It was proved to give phonon modes of GNRs compatible with density functional theory calculations [40]. The calculated force constants are used as the phonon Hamiltonian. The complex many-body interactions (phonon–phonon and electron–phonon interactions), which are only important at very high temperatures, are neglected here. By following the procedures described previously, we calculate the phonon Green's function and then phonon transmission, and finally compute thermal conductance as a function of temperature.

The calculated scaled thermal conductance at 300 K versus width for GNRs and CNTs are presented in Fig. 3.8a. It is very interesting that thermal conductance of GNRs and CNTs show quite different shape and width dependence. Specifically, ACNTs have nearly the same scaled thermal conductance as ZCNTs, consistent with previous theoretical results [41]. In addition, the scaled thermal conductance of CNTs weakly depends on their circumference. Based on these facts, we established that the scaled thermal conductance of CNTs is insensitive to the change of their chirality and ribbon width. In contrast, significant size effect and anisotropy of thermal conductance are observed in GNRs. There, the scaled thermal conductance decreases rapidly as the ribbon width increases for narrow GNRs (width smaller than 2 nm), showing strong quantum confinement effects. While for wider GNRs, the size dependence becomes weaker but remains observable for all the GNRs considered by our study. More importantly, ZGNRs have significantly larger scaled thermal conductance than AGNRs, indicating a strong anisotropic thermal transport in GNRs.

The appearance of anisotropy in thermal conductance is really unexpected, if considering the fact that both graphene (i.e., the bulk counterpart of GNRs) and CNTs (i.e., rolled-up GNRs) are isotropic in thermal conductance. This observation raises many interesting and important questions. What is the origin of the anisotropy? Is this a general phenomenon? How does the anisotropy evolve as increasing the ribbon width? Can it be used for thermal conduction tuning? For answering these questions, let us discuss the anisotropic thermal conductance of GNRs in more details.

We define an anisotropy factor for the thermal conductance of GNRs as

$$\eta = [(\sigma/S)_{\mathrm{ZGNR}}/(\sigma/S)_{\mathrm{AGNR}}] - 1 \tag{3.66}$$

to give a quantitative description of the anisotropic thermal conductance. As shown in Fig. 3.8b, narrower GNRs generally exhibit stronger anisotropy, though some irregular variation of the anisotropy factor is shown up when W is less than 2 nm. The room temperature anisotropy factor changes irregularly from 31%

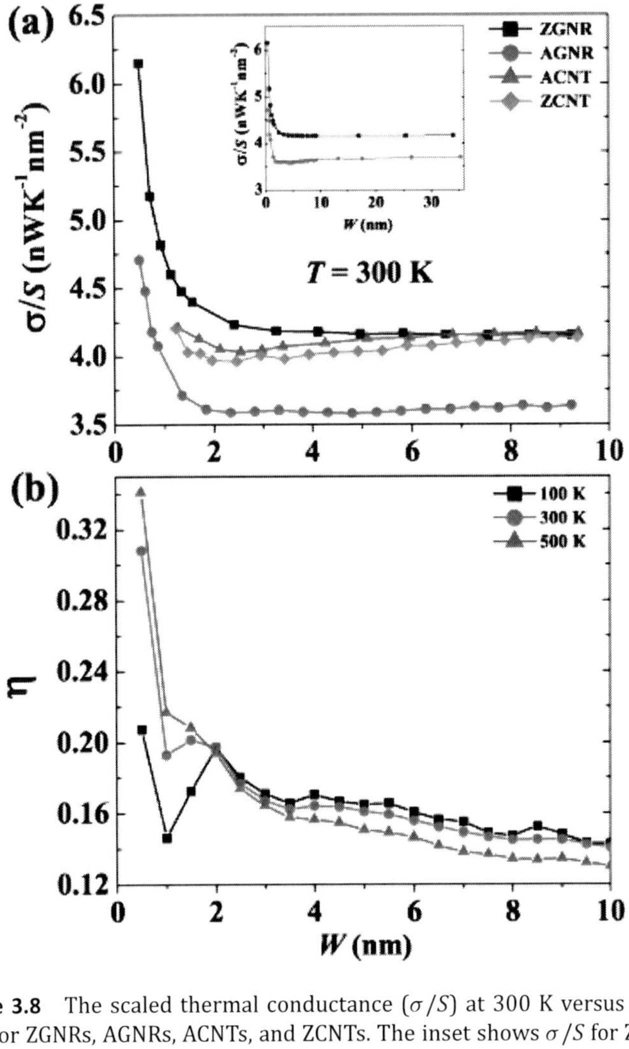

Figure 3.8 The scaled thermal conductance (σ/S) at 300 K versus width (W) for ZGNRs, AGNRs, ACNTs, and ZCNTs. The inset shows σ/S for ZGNRs and AGNRs with the width varying from 0.5 to 35 nm. (b) The anisotropy factor (η) versus width (W) at 100 K, 300 K and 500 K. The lines are drawn to guide the eyes. Reprinted with permission from Ref. [38]. Copyright 2009, AIP Publishing LLC.

to 20% when W varies from 0.5 nm to 2 nm, and decreases monotonously to 14% as W increases to 10 nm. Since thermal conductance is isotropic in graphene sheets [42], we expect the anisotropy factor of GNRs will eventually decease to zero when W is large enough. But what is the critical width at which the anisotropy of thermal conductance disappears? To determine the critical width, we considered wider GNRs with W up to 35 nm. Their scaled thermal conductance, as shown in the inset of Fig. 3.8a, varies very slowly with W. Same is for the anisotropy factor. The room temperature anisotropy factor of 35 nm wide GNR still has a value of 13%. Obviously, a direct calculation of thermal conductance at critical width using the NEGF method is beyond our computational capability. So we apply linear regression to fit the data from 4 to 35 nm for ZGNRs and AGNRs, and find that the room temperature anisotropy may disappear when $W \sim 140$ nm.

The temperature-dependence of the anisotropy factor is also investigated as shown in Fig. 3.8b. Varying temperature from 100 K to 500 K only slightly affects the anisotropy factor except for very narrow GNRs. While the anisotropy factor will decrease rapidly when the temperature decreases from 100 K (the data is not shown here). An extreme case is that at nearly zero temperature the anisotropy factor will be zero because thermal conductance of GNRs, irrespective of edge shapes, are all $4\sigma_0$, where σ_0 is the quantum of thermal conductance.

We have analyzed the size-dependence and temperature-dependence of the anisotropy. What is the underlying mechanism for this anisotropy? Similar anisotropy is found in the thermal conductance of silicon nanowires (SiNWs) [43]. There, the anisotropic phonon structure of bulk silicon is proposed to be the reason. However, same mechanism does not apply for GNRs, because their bulk counterpart, graphene, is isotropic in thermal conductance. A MD calculation on thermal conductance of GNRs attributes the anisotropy to different phonon scattering at edges. This is obviously not the case here, because there is not boundary scattering for periodic structure studied here. We showed that this anisotropy of thermal conductance appears even without phonon scattering. Our results suggest that it is the different boundary conditions at edges instead of bulk property or phonon scattering that gives rise to the anisotropy in thermal conductance.

Boundary condition affects thermal transport in a very complicated way due to the collective nature of lattice vibrations. Every phonon mode will be influenced if the boundary condition varies. To analyze the influence on thermal conductance, which is a weighted integration of the contribution of all the phonon modes excited, we look at the whole phonon transmission spectra rather than focus on individual phonon modes. For comparison we select two GNRs, 16-ZGNR and 28-AGNR, which share the same ribbon width ($W \sim 3.3$ nm) but different edge shape. The former one gives higher thermal conductance than the latter one. The influence of the different boundary condition on their phonon transmission, as shown in Fig. 3.9a, is strongly frequency-dependent. Negligible difference in phonon transmission is observed at low-frequency region ($\omega < 100$ cm^{-1}), explaining the nearly zero anisotropy at very low temperatures. While phonon transmission of 16-ZGNR is obviously higher than that of 28-AGNR at high-frequency region (especially at 400$\sim$600 cm^{-1} and 1400$\sim$1650 cm^{-1}). The difference in thermal conductance of the two GNRs is obvious at high temperatures, as shown in Fig. 3.9b.

Next, we artificially modify the boundary conditions of GNRs to further understand its effect on the anisotropy in thermal conductance. This is achieved by setting the mass of edge atoms to infinitely large (10^7 atomic mass units in the calculation) so as to impose fixed boundary condition on the above two GNRs (i.e., 16-ZGNR and 28-AGNR). As shown in the inset of Fig. 3.9b, due to the fix of edge atoms in GNRs, the scaled thermal conductance decreases for both GNRs. Differently, the decrease is relatively smaller in 28-AGNR than in 16-ZGNR, indicating that the boundary condition imposed by armchair edge is more like fixed boundary condition. The anisotropy becomes smaller but does not disappear after imposing fixed boundary condition. This, on the other hand, shows that the anisotropy is robust even under this extreme condition, which is substantially different from the free or periodic boundary condition generally used.

Finally we show that the boundary effects on thermal conductance can be qualitatively compared for different edge shapes.

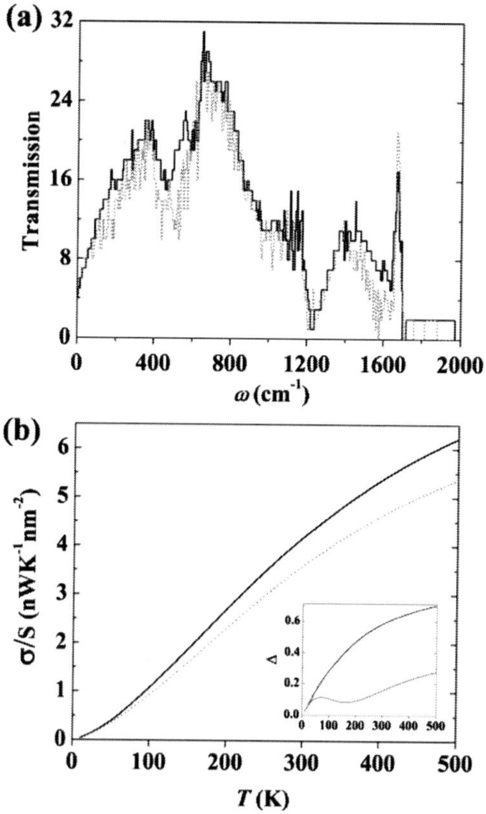

Figure 3.9 (a) Transmission function versus phonon frequency (ω) and (b) the scaled thermal conductance (σ/S) versus temperature (T) for 16-ZGNR and 28-AGNR. The inset shows the decrease of scaled thermal conductance (Δ) induced by fixing edge atoms. Reprinted with permission from Ref. [38]. Copyright 2009, AIP Publishing LLC.

Graphene, the bulk counterpart of GNRs, is isotropic in thermal conductance. Cutting graphene into GNRs introduces edges, where the bonding environment is different from in the bulk. Different edge shapes show different transport ability. Generally, if the edge-induced change in the bonding strength is smaller along the transport direction, phonons of quasi-1D transport systems would become more delocalized, thus phonon transmission and thermal conductance would be higher. This suggests that by analyzing the bonding environment at edges, we can qualitatively compare the delocalization degree of phonons between quasi-1D transport systems with different boundary conditions. The approach is simple and can explain the different scaled thermal conductance observed for GNRs and CNTs. In GNRs, the bonds that are cut off due to the formation of edges are perpendicular to the transport direction for zigzag edges, while they have components along the transport direction for armchair edges. According to our approach, ZGNRs would give higher scaled thermal conductance than AGNRs. If rolling up GNRs into CNTs, the bonding configuration of the bulk (i.e., graphene) is preserved in the whole system. We would expect that the scaled thermal conductance of CNTs is close to that of ZGNRs, and weakly depends on the chirality and diameter of CNTs. All these expectations given by this approach are consistent with the results of our calculations.

We have shown that thermal conductance is anisotropic in GNRs due to the different boundary conditions at edges. The appearance of anisotropy is independent of phonon scattering and bulk phonon structure. Actually, the anisotropy found here is not limited to GNRs, but is expected to also exist in other materials. For instance, similar anisotropy is found in boron nitride nanoribbons as revealed by our calculations [44]. Our study demonstrates that by controlling bonding environments at edges, the delocalization degree of phonons, and thus the thermal conductance, can be tuned. This conclusion generally works for any materials. Another important observation is that the boundary effects on phonon systems are very long range. Using GNRs as an example, the room temperature anisotropy in thermal conductance is significantly large (13%) even for GNRs as wide as 35 nm and would only disappear at $W > 140$ nm. The difference in thermal conductance, caused by different edge shapes, can be very large for wide GNRs. All these indicate that thermal conductance can be tuned in a wide range by edge control.

3.4.2 *Origin of High Thermal Conductivity in Graphene*

Graphene possesses extremely high thermal conductivity, as demonstrated experimentally and theoretically. Thermal conductivity (κ) is related to thermal conductance (σ) by

$$\kappa = \sigma L/S, \tag{3.67}$$

where L and S are the length and the cross sectional area of the sample, respectively. Once L becomes long enough, thermal transport enters into the diffusive region and $\sigma \propto 1/L$ Approximately, $\sigma L \approx \sigma_b l$, where σ_b is the ballistic thermal conductance and l is the effective phonon mean free path. Then we have

$$\kappa \approx \sigma_b l/S. \tag{3.68}$$

The extremely high thermal conductivity may come from the extraordinary high ballistic thermal conductance. Or it may be caused by the superiorly long phonon mean free path. What is the case in reality? It is important to answer this question, because these two possibilities point two radically different directions for the applications of graphene.

3.4.2.1 Ballistic thermal conductance of graphene

In the following, we will study the ballistic thermal conductance of graphene, which is actually the upper limit to the thermal conductance. As far as we know, two approaches are available to calculate ballistic thermal conductance of 2D periodic systems (graphene here). The first method uses the Landauer formula that employs ballistic phonon transmission function calculated directly from phonon dispersion by counting transport channels. The second approach does not directly deal with the quasi-2D system. Instead, thermal transport in quasi-1D systems of different widths is studied. After knowing the size-dependence, thermal-transport properties of the quasi-2D system are obtained by extrapolating results of the finite-width systems to the infinite-width limit. Both approaches are equivalent to each other and predict same results, as we will demonstrate below.

The Landauer formula is applicable not only for 1D systems as we have showed, but also for 2D and 3D systems. For a 2D system, ballistic thermal conductance per width contributed by phonons as a function of temperature ($\sigma_{\text{ballistic}}(T)/w$) is related to phonon dispersion ($\omega_m(\mathbf{k})$ in which m is the index of phonon bands and k

denotes the phonon wave vector) by

$$\sigma_{\text{ballistic}}(T)/w = \sum_{m} \int_{v_x>0} \frac{dk_x}{2\pi} \int \frac{dk_y}{2\pi} \hbar\omega_{\text{m}}(\mathbf{k}) v_{xm}(\mathbf{k}) \frac{\partial f(\hbar\omega_{\text{m}}(\mathbf{k}), T)}{\partial T}, \tag{3.69}$$

where $v_{xm}(\mathbf{k}) = \partial\omega_{\text{m}}(\mathbf{k})/\partial k_x$ is the phonon group velocity along the direction (i.e., the transport direction) and f is the Bose–Einstein distribution function [42].

Similarly, ballistic thermal conductance per width contributed by electrons as a function of temperature ($\sigma^e_{\text{ballistic}}(T)/w$) is related to the energy dispersion of electrons ($\varepsilon_{\text{m}}(\mathbf{k})$, in which m is the index of electronic bands and $\mathbf{k}$ denotes the electron wave vector) by

$$\sigma^e_{\text{ballistic}}(T)/w = 2\sum_{m} \int_{v_x>0} \frac{dk_x}{2\pi} \int \frac{dk_y}{2\pi} [\varepsilon_{\text{m}}(\mathbf{k}) - \mu] v_{xm}(\mathbf{k}) \frac{\partial f_{\text{FD}}(\varepsilon_{\text{m}}(\mathbf{k}), T)}{\partial T}, \tag{3.70}$$

where the extra factor of 2 accounts for the spin degeneracy of each band, μ is the Fermi energy, $v_{xm}(\mathbf{k}) = \partial\varepsilon_m(\mathbf{k})/\partial k_x$ is the electron group velocity along the x direction (i.e., the transport direction) and f_{FD} is the Fermi–Dirac distribution function [42]. The ballistic thermal conductance contributed by electrons and the ballistic electronic conductance satisfy the Wiedemann–Frantz law. Both of them will change if the Fermi-level of grapheme is shifted. The electronic thermal conductance is important only at low temperatures and discussed in detail in Ref. [42]. Here we focus on the thermal conductance contributed by phonons.

Once we get the phonon dispersion of graphene, ballistic thermal conductance can be easily calculated by the Landauer formula. The phonon dispersion can be calculated using the empirical potential as in Ref. [42] or using the more accurate first-principles methods. Fig. 3.10 presents the phonon dispersion of graphene obtained by DFT calculations. Each unit cell of graphene contains two carbon atoms, thus the phonon dispersion of graphene includes six phonon branches: (i) out-of-plane acoustic (ZA) and out-of-plane optical (ZO) phonons, which correspond to vibrations along the z direction (i.e., perpendicular to the graphene plane); (ii) transverse acoustic (TA) and transverse optical (TO) phonons, which correspond to in-plane transverse vibrations; (iii) longitudinal acoustic (LA) and

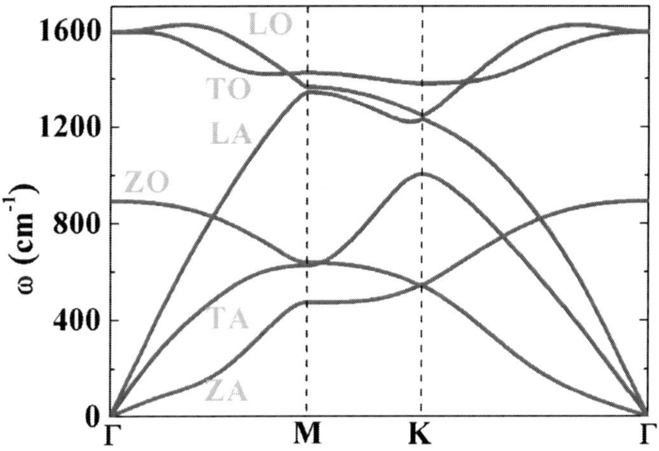

Figure 3.10 Phonon dispersion of graphene computed by density functional theory (DFT). The calculation uses norm conserving pseudo potential, plane wave basis with an energy cut off of 40 Hartree, and Troullier-Martins density functional in local density approximation (LDA) as implemented in the ABINIT code. Reprinted from Ref. [45], Copyright (2009), with permission from Elsevier.

longitudinal optical (LO) phonons, which correspond to in-plane longitudinal vibrations.

At long wavelengths the LA and TA phonon modes have a linear dispersion. In contrast, due to rotational symmetry the flexural ZA phonon mode obeys a quadratic dispersion in the vicinity of the Γ point of the Brillouin zone, which is a characteristic feature of layered crystals [46]. Due to the existence of this quadratic dispersion, at low temperatures thermal conductance of graphene is dominated by the $T^{1.5}$ contribution of the ZA phonon mode, in contrast to the T^2 contribution of the linear LA and TA phonon modes [41]. Note that the numerical calculations may slightly break the rotation symmetry, which can cause the dispersion of the ZA mode deviate from quadratic. Therefore, numerically reproducing the quadratic dispersion is difficult. In fact, our calculated ZA branch is linear close to the Γ point. This inaccuracy in phonon dispersion changes the temperature dependence of thermal conductance at low-temperature region (less than 50 K). Nevertheless, the temperature dependence of

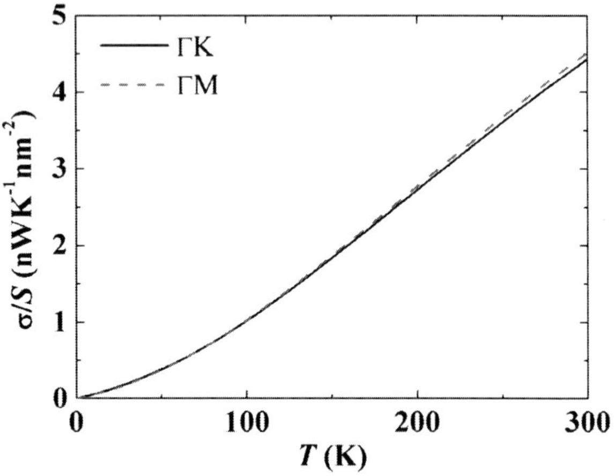

Figure 3.11 The ballistic thermal conductance per unit area (σ/S) of graphene as a function of temperature (T), calculated by the Landauer formula, which employs the phonon dispersion shown in Fig. 3.10. The transport direction is chosen to be along the ΓM or the ΓK direction. The cross sectional area $S = w\delta$, where w is the width of the graphene and $\delta =$ 0.335 nm is chosen as the layer separation in graphite.

thermal conductance at higher temperatures is almost unaffected. Moreover, the induced inaccuracy in the calculated ballistic thermal conductance is estimated to be less than 10% [41].

We calculated the ballistic thermal conductance of graphene applying the Landauer formula using the computed phonon dispersion (shown in Fig. 3.10). The calculated scaled ballistic thermal conductance is presented in Fig. 3.11. Selecting different transport directions in graphene leads to only slight difference in the calculated scaled thermal conductance, consistent with previous results [47]. This directly proves that thermal transport is isotropic in graphene. The ballistic thermal conductance of graphene grows as increasing temperature. At room temperature, a value of 4.4~4.5 nW/K/nm^2 is obtained.

Alternatively, we study the ballistic thermal conductance of graphene indirectly by investigating thermal conduction of GNRs and CNTs. The scaled ballistic thermal conductance of GNRs and CNTs with various widths is presented in Fig. 3.8a. From those results, we extrapolate that the scaled ballistic thermal conductance of graphene is $\sim$4.2 nW/K/nm^2 at room temperature. The two approaches give essentially the same result. The slight difference in the predicted thermal conductance arises from the fact that the energy potential of graphene is described by different approaches: the former calculation uses first-principles method and the latter uses empirical potential.

The two separated calculations consistently predict that the scaled ballistic thermal conductance is $\sim$4 nW/K/nm^2 at room temperature, consistent with previous study [41]. As a comparison, the scaled room temperature ballistic thermal conductance of pristine silicon nanowires (SiNWs) is $\sim$1 nW/K/nm^2 [43], of the same order as that of graphene. This cannot explain the three orders of magnitude difference in their thermal conductivity measured by experiments. Our results demonstrate that it is the very long phonon mean free path, instead of the extremely high ballistic thermal conductance, that gives rise to the extraordinary high thermal conductivity of graphene.

3.4.2.2 Long phonon mean free path in graphene

In last subsection, we theoretically demonstrate that the extraordinary high thermal conductivity of graphene comes from its superiorly long phonon mean free path. Thermal measurements using a noncontact technique based on micro-Raman spectroscopy estimate that the phonon mean free path of graphene is around 775 nm at room temperature [30]. An estimation of thermal conductivity, employing the Eq. 3.68 using the calculated ballistic thermal conductance and the experimental phonon mean free path, gives a value about 3000 W/m/K, consistent with previous experimental measurements.

What is the underlying mechanism of the long phonon mean free path in graphene? First, the many body interactions (electron–phonon and phonon–phonon interactions) are weaker in 2D systems than 3D systems, because the requirement of simultaneously satisfying both the momentum and energy conservation laws exerts stronger restrictions on scattering events in systems of lower dimensions. Second, structural distortion has minor effects on thermal transport in graphene. Our recent study finds that even a strong structural deformation can be viewed as a perturbation of the transport of most low-frequency phonons in CNTs [48]. We expect similar feature in graphene because of the similar sp^2 bonding configuration that is well preserved even under severe deformation. Third, introducing an isolated structural defect (such as vacancy or substitution) in graphene-like structures almost does not affect the transport of low-frequency acoustic phonons and only slightly scatters high-frequency optical phonons [49]. Extended defects or isolated defects of high concentration may significantly lower thermal conductance of graphene [50]. However, the defect formation energy is very high in graphene because of the strong bonding configuration. Last, but not least, interface scattering is weak in all-graphene transport systems, characterized by the very low thermal contact resistance [51]. In short summary, benefitting from unique 2D plane structure and strong sp^2 bonding configuration, phonons in graphene experience weak many-body interactions, structural distortion induced scatterings, defect scatterings and interface scatterings, all of which are much weaker than in 3D materials (such as bulk silicon). This gives rise to very long phonon mean free path in graphene.

3.4.3 *Thermal Transport in Graphene-Based Devices*

The finding of graphene has inspired enormous research interest from fundamental science to emerging technologies. Using graphene as the base material to replace silicon for future nanoeletronics is one of the most fascinating and promising research directions. In graphene-based nanoelectronics, there are two major building blocks to achieve transistor action. One is graphene junctions built by GNRs. As shown in Fig. 3.12, the patterned GNRs combined with controlled doping and manipulation of edge termination can be used for building electronic devices, like metal-semiconductor junctions, *p–n* junctions, heterojunctions, and field effect transistors (FETs) [52]. These GNR-based devices are believed to exhibit superior performance over the conventional silicon-based devices. For instance, first-principles calculations reveal that GNR-FETs show outstanding performance, characterized by their very large ON/OFF ratio of the order of $10^3 \sim 10^4$ [52]. The other major building block is graphene quantum dots (QDs) [53]. Using etched QDs (shown in

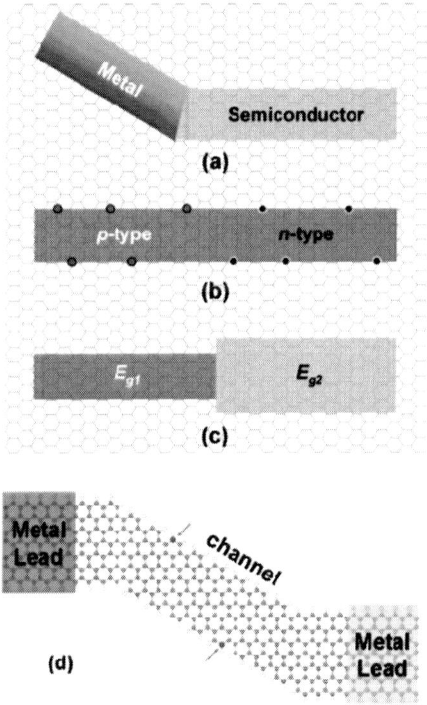

Figure 3.12 Schematics of three device building blocks: (a) a metal-semiconductor junction between an armchair and a zigzag GNR, (b) a *p*-*n* junction between two zigzag GNRs with different edge doping, and (c) a heterojunction between two zigzag GNRs of different width (band gap). (d) Schematics of a GNR-FET, made from one zigzag semiconductor GNR channel and two armchair metallic GNR leads connected to two external metal electrodes. Reprinted with permission from Ref. [52]. Copyright 2007 American Chemical Society.

Fig. 3.13), graphene-based single-electron transistors are realized as demonstrated by a recent experiment [54].

One of the most attractive features of graphene-based electronics is that everything in electronics including conducting channels, quantum dots, barriers, electrodes, and interconnects can be patterned in a single graphene plane. For instance, the flip-flop circuit for high-frequency signal generation, as an example integrated circuit, can find its graphene-based counterpart as depicted in Fig. 3.14. As a consequence, atomic perfect interfaces are naturally

Figure 3.13 Graphene devices. (Left) Scanning electron micrograph of a graphene-based single-electron transistor. The dark areas are gaps in the photoresist mask where graphene is removed by plasma etching. The small quantum dot in the center is connected to contact regions through narrow constrictions, and there are four side gates. (Right) Schematic showing a hypothetical graphene-based single-electron transistor based on a quantum dot that contains only a few benzene rings. Reprinted from Ref. [54] with kind permission from Prof. Sir Andre Konstantin Geim, University of Manchester, UK.

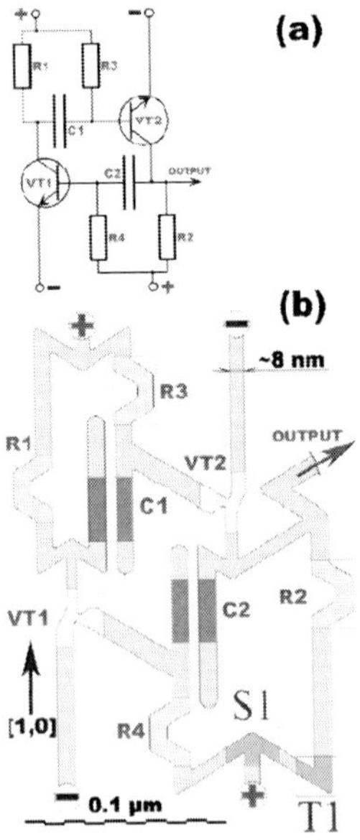

Figure 3.14 (a) Flip-flop circuit without wire crossings. (b) Its possible GNR-based counterpart patterned from a single graphene sheet. To keep this drawing compact, the capacitors C1 and C2 are not drawn to scale. Reprinted with permission from Ref. [55]. Copyright 2007 American Chemical Society.

achieved and the contact resistances between the devices and electrodes are very low. More importantly, the plane-patterned graphene-based electronics can benefit from the advantages of the existing planar technologies.

In ever-smaller integrated circuits, the heat dissipation becomes more and more serious due to the increased heat generation per unit area and the reduced thermal conduction caused by increasingly strong boundary and interface scattering. Therefore, thermal management has become a critical issue for future development of (graphene-based) nanoelectronics. At present, most research focuses on the electronic transport properties of graphene-based nanodevices. In contrast, much less is known about their thermal-transport properties, though such kind of knowledge and understanding is urgently needed for practical applications. We investigate thermal-transport properties of graphene-based devices by the NEGF method. More details of this study can be found in Ref. [51].

We focus on two types of nanodevices: graphene junctions and QDs, both of which are major building blocks of graphene-based electronics. It is well known that electronic transport in graphene junctions and QDs are strongly dependent on their structural characteristics, such as the contact geometry, width, edge shape, connection angle, and so on. To the best of our knowledge, the corresponding knowledge on the influence of these structural freedoms on thermal-transport properties is, though of highly importance for designing graphene devices, still missing. Several theoretical and experimental works have been done on thermal transport in 2D graphene sheet or individual GNRs. However, the realistic graphene-based nanodevices have more complex structures and thus their thermal-transport properties cannot be deduced solely from those of the regular graphene sheet or GNRs. This motivates us to systematically study the influence of structural characteristics on thermal transport in graphene junctions and QDs.

Here thermal conductance contributed by electrons is not considered, because of two reasons: (i) this contribution can be evaluated by the well known electronic transport properties, because thermal conductance and electronic conductance contributed by electron satisfy the Wiedemann–Frantz law; and (ii) thermal conductance of narrow GNRs is mostly dominated by phonons, since all sub–10 nm GNRs are semiconducting as shown by recent experiment [56]. In our NEGF calculations, we neglect the complex many-body interactions (i.e., phonon–phonon and electron–phonon interactions), which induce very few phonon scatterings because of the extremely long phonon mean free path in graphene (around 775 nm at room temperature [30]), much longer than the length of the central part of our calculated transport systems. With such approximation that does not affect our final results, the quantum thermal-transport problem can be solved exactly by the NEGF method. We calculated the phonon Hamiltonian (i.e., force constant matrix) by the second-generation reactive empirical bond order potential [39], which has been demonstrated to be able to predict accurate phonon structures for graphene-based systems [40]. Phonon transmission and phonon local density of states (LDOS) are given by the NEGF calculations. Thermal conductance as a function of temperature is computed by the Landauer formula.

3.4.3.1 Contact geometry

Connecting two GNRs together can form various single-interface graphene junctions that could be used for basic device building blocks (Fig. 3.12) [52]. Electronic transport in such junctions is well known to be strongly dependent on the geometrical structure in the contact region. Here we study effects of contact geometry on thermal transport, by considering various single-interface junctions that share the same composed GNRs but have different structure close to the interface. Though contact geometries are distinctly different, their thermal conductance only shows slight variances. Thermal transport is insensitive to the changes in contact geometry, in contrast to electronic transport, which shows strong dependence on contact geometry. The distinct feature of electronic and thermal transport can be explained from two aspects: (1) the typical length scales (e.g., wavelength and mean free path) of phonons and electrons are different; and (2) only electrons near the Fermi level contribute to electronic conductance, while all phonon modes that are populated (or excited) contribute to thermal conductance. Since different contact geometries give rise to negligible difference in thermal conductance, without losing generality we will investigate graphene junctions that have sharp interfaces.

3.4.3.2 Ribbon width

Three types of graphene-based devices are considered: single-interface junctions, double-interface junctions and QDs, which are composed of GNRs as shown in Fig. 3.15a. A simplified model of QD is used here, characterized by a center part connected by two narrow constrictions on the left and right. More realistic models of QD will be discussed later. For these graphene-based nanodevices, one of most important structural parameters is the ribbon width. In this section, we study the influence of ribbon width on thermal transport. The edge of composed GNRs is limited to be zigzag here, and the influence of edge shape on thermal transport will be studied by comparing with armchair edges. The two GNRs,

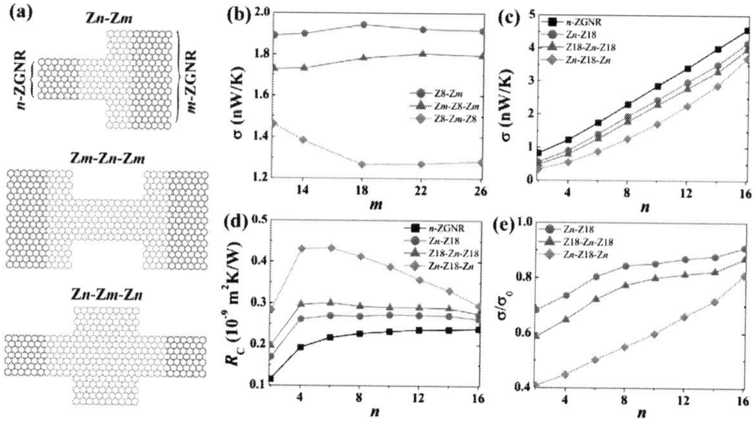

Figure 3.15 Schematics of graphene junctions and quantum dots formed by *n*-ZGNR and *m*-ZGNR ($n < m$): single-interface junction "Z*n*-Z*m*", double-interface junction "Z*m*-Z*n*-Z*m*" and quantum dot "Z*n*-Z*m*-Z*n*". (b) Thermal conductance (σ) at 300 K as a function of structure parameter *m* for "Z8-Z*m*" (circle), "Z*m*-Z8-Z*m*" (triangle) and "Z8-Z*m*-Z8" (diamond), (c) Thermal conductance (σ), (d) thermal contact resistance (R_C), and (e) thermal conductance ratio (σ/σ_0) at 300 K as a function of structure parameter *n* for pristine *n*-ZGNR (square), "Z*n*-Z18" (circle), "Z18-Z*n*-Z18" (triangle) and "Z*n*-Z18-Z*n*" (diamond) [51].

n-ZGNR and m-ZGNR ($n < m$ is assumed without loss of generality), are selected to construct single-interface junction, double-interface junction and QD, which are labeled as "Zn-Zm" "Zm-Zn-Zm" "Zn-Zm-Zn" for convenience. Then n and m representatively correspond to the widths of the narrow and wide parts of the structures. The length of the central GNR is fixed to be as long as shown in Fig. 3.15a. Our calculations show that further increasing its length has negligible influence on the calculated thermal conductance.

Previous studies reveal that varying the ribbon width strongly affects the performance of graphene-based electronic nanodevices. For instance, the room temperature on/off ratio of GNR FET induced by gate voltage increases exponentially as the ribbon width decreases [56]. We find that thermal-transport behavior has a more complicated dependence on the ribbon width. Varying the width of wide and narrow part affects thermal transport in a distinctly different way. As presented in Fig. 3.15b, thermal conductance of the single-interface and double-interface junctions is insensitive the width change in the wide part. This is also the case for the QD structure except when the width of the wide center part close to that of the two narrow neighboring parts. The wide part in the QD structure acts as a protrusion along the GNR. Once the protrusion disappears, thermal conductance will show a noticeable increase. In contrast, thermal conductance of all the three types of graphene structures strongly dependent on the width of the narrow part. Thermal conductance of the single-interface and double-interface junctions is almost linearly proportional to the width of the narrow part, while that of the QD approximately shows a quadratic dependence, as presented in Fig. 3.15b. Since the number of phonon transport channels is limited by the narrow part, decreasing its width will remarkably reduce thermal conductance of the whole system.

Due to interface scattering, thermal conductance of the three graphene structures is always lower than that of the corresponding pristine narrow GNR (i.e., n-ZGNR) as demonstrated in Fig. 3.15b. Thermal conductance of GNRs increases approximately linearly with the ribbon width. Due to this feature of strong width-dependence, thermal conductance itself is not suitable to measure the thermal-transport ability of a transport system. Instead, to quantify the influence of interface scattering, we define two quantities for the graphene structures: thermal contact resistance $R_c = W\delta/\sigma$, which equals to the temperature difference between the two thermal leads per unit heat flux [57], and thermal conductance ratio σ/σ_0, where W and σ_0 are, respectively, the width and the ballistic thermal conductance of the pristine narrower GNR (i.e., n-ZGNR) and $\delta = 0.335$ nm is chosen as the layer separation in graphite. The calculated R_c and σ/σ_0 at room temperature, is shown in Figs. 3.15c and 3.15d, respectively, for the graphene structures with different width of the narrow part.

As presented in Fig. 3.15d, when n (corresponds to the width of the narrow part) increases, σ/σ_0 of graphene junctions grows at small n (less than 10) and becomes (almost) saturated, while that of QDs exhibits a nearly linear increase. Larger n always gives higher σ/σ_0 for all the three graphene structures, implying that thermal conduction reduction induced by interface scattering is relatively weaker as the narrow part gets wider. When the narrow part is about 2 nm (i.e., $n = 10$), σ/σ_0 at 300 K are about 85%, 80%, and 60% for the single-interface junction, double-interface junction, and QD respectively. To some extent, these predicted values quantify thermal-transport ability of realistic graphene nanodevices, since thermal conductance of the graphene structures varies slightly when the width of wide part changes. Our results indicate that thermal-transport ability of single-interface junction is close to that of double-interface junctions, but obviously better than that of the QD structure.

As shown in Fig. 3.15c, thermal contact resistance of the single-interface and double-interface junctions is almost width-independent, while that of the QD structure gradually decreases with the width of the narrow part. The irregularity appears at $n = 2$, presumably caused by predominant edge effects. Thermal contact resistance of graphene junctions is insensitive to the variance in ribbon width. Benefitting from this feature, we can extrapolate that in realistic graphene based nano-junctions, the room temperature thermal contact resistance between ZGNRs and graphene is around 0.3×10^{-9} m^{2}K/W. This value is much lower than those between graphene and silicon dioxide (5.6×10^{-9}–1.2×10^{-8} m^{2}K/W [58]).

It would be interesting to compare thermal contact resistance among the three types of graphene structure. The single-interface junction always gives lower thermal contact resistance than its double-interface counterparts. This is easy to understand because an extra interface exerts additional phonon scattering. The double-interface junction and the QD structure share two similar interfaces, however, their thermal contact resistance differs significantly. The QD has obviously larger thermal contact resistance than the double-interface junction. This is unexpected in the classical picture. When thermal resistances are connected in series, they are additive. If using the pristine *n*-ZGNR as reference system, introducing an interface produces additional thermal contact resistance, which is given by the thermal contact resistance difference between the single-interface junction and the reference system. Thus, an interface in fact acts as a thermal resistor. Two of them are placed in series in the double-interface junction and QD. The increase of thermal contact resistance with respect to the reference system in the double-interface junction/QD, however, is not equal to but lower/higher than two times that in the single-interface junction. The conventional additive law fails, since quantum interference effect plays an important role in transport [4]. Transport properties cannot be determined simply by the number of interfaces.

We further analyze the phonon transmission function to understand the differences in thermal conductance and thermal contact resistance. Transport systems composed of 8-ZGNR and 18-ZGNR is selected for our following discussion without losing generality. As shown in Fig. 3.16a, phonon transmission function of the single-interface junction is slightly that of the narrow GNR, but largely lower than that of the wide GNR (data not shown here). This result indicates that phonons from the narrow GNR experience little scattering when transporting across the interface into the wide GNR, while phonons from the wide GNR are largely scattered as they transport into the narrow GNR. In other words, the average transmission per mode is strongly direction-dependent, due to the asymmetry structure of the single-interface junction. This can qualitatively explain the different transport properties in the three types of graphene structures. Comparing with the single interface junction, the double-interface junction and the QD have one more

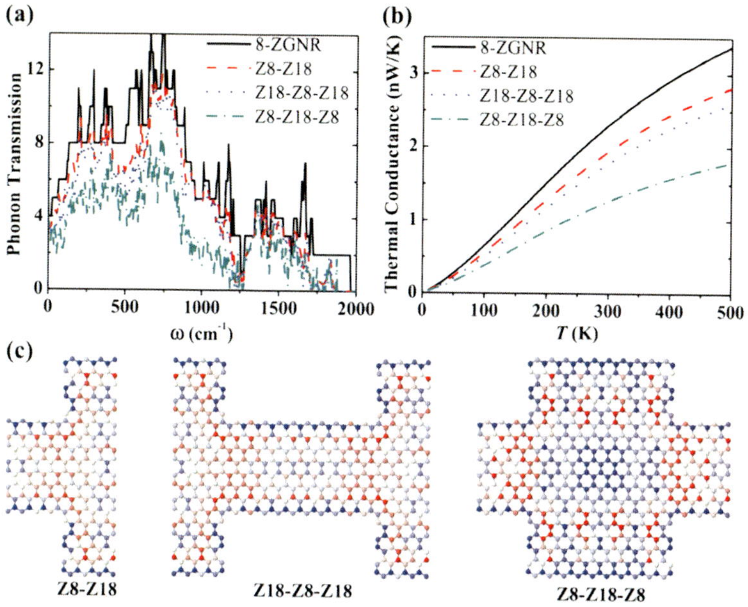

Figure 3.16 (a) Phonon transmission as a function of frequency ω and (b) thermal conductance as a function of temperature T for pristine 8-ZGNR (black solid line), "Z8-Z18" (red dashed line), "Z18-Z8-Z18" (blue dotted line) and "Z8-Z18-Z8" (cyan dash-dotted line), (c) Schematics showing the phonon local density of state at $\omega = 900\ \text{cm}^{-1}$ for "Z8-Z18", "Z18-Z8-Z18" and "Z8-Z18-Z8" structures. Red (blue) color represents the largest (smallest) value [51].

interface. To across this additional interface, phonons in the double-interface junction transmit from the narrow GNR into the wide GNR, different from those in the QD. Due to the strong direction-dependence of interface scattering, phonon scattering introduced by the second interface is small in the double-interface junction but large in the QD. As evident, phonon transmission of the double-interface junction is slightly lower than that of the single-interface junction, but apparently larger than that of the QD in almost all the frequency range especially at low-frequency region (Fig. 3.16a). Consequently, thermal conductance of graphene junctions is significantly higher than that of the QD in the whole temperature range (Fig. 3.16b).

The phonon LDOS characterizes the distribution of phonons in real space and can provide more detailed information on thermal transport. To see the difference in thermal-transport behavior, we plot phonon LDOS for the three systems at $\omega = 900\ \text{cm}^{-1}$ where their phonon transmission is noticeably different. As shown in Fig. 3.16c, the atoms with large LDOS is mainly distributed in the inner part of the two graphene junctions, indicating that phonons are more likely to appear in that region. Therefore, straight channels along the transport direction are available for phonons and the reflection probability is small. Differently, large LDOS appears in the inner part of the narrow section and the protruding part of the wide section in the QD. The transport trajectory is not straight, giving rise to large scattering probability. This rationalizes why the interface scattering is much stronger in the QD than in the junctions.

3.4.3.3 Edge shape

It is well known that the electronic transport properties of AGNRs are significantly different from those of ZGNRs [18]. We showed that the edge shape can also noticeably affect thermal-transport properties of pristine GNRs [38]. However, its influence on thermal transport in more complex graphene-based transport systems is still unknown. To explore edge effects, we further considered graphene structures composed of AGNRs. Our results show that thermal-transport characteristics of the graphene structures with armchair edges are very similar to those with zigzag edges, but the structures with armchair edges have higher thermal contact resistance and lower thermal conductance ratio. Thermal contact resistance of the junctions with armchair edges, which approximately equals to that between AGNRs and graphene, is $0.4 \times 10^{-9}\ \text{m}^2\text{K/W}$ at 300 K, about 30% higher than the case of zigzag edge. Thermal conductance ratios of the single-interface junction, the double-interface junction, and the QD are 77%, 76%, and 54%, respectively, at 300 K when the width of the narrow part is about 2 nm. These values are sizably lower than the corresponding values of structures with zigzag edges. Our calculations reveal that zigzag edge shows better thermal-transport ability than armchair edge in graphene-based nanodevices.

3.4.3.4 Connection angle

Experimentally graphene junctions formed by GNRs with different connection angles can be produced [56, 59]. We investigate the influence of connection angle on thermal transport in single-interface graphene junctions. For convenience, we label these junctions according to the edge shape and connection angle of the composed GNRs as shown in Fig. 3.17a. For instance, "ZA90" denotes a single-interface junction constructed by connecting ZGNR to AGNR with the connection angle of 90°. Their thermal conductance as a function of temperature is presented in Fig. 3.17b. Generally, the larger the connection angle, the higher thermal conductance. As the temperature increases, the change in the connection angle of graphene junctions gives rise to larger differences in thermal conductance. To evaluate the influence of the connection angle, we further compare their thermal conductance ratio (σ/σ_0) that is comparable between different materials and transport systems. As shown in Fig. 3.17c, σ/σ_0 only slightly increases from 100 K to 500 K. Larger connection angle leads to higher σ/σ_0. For instance, σ/σ_0 of the "ZZ120" junction is 78% at room temperature, 1.7 times that of "ZZ60". Our results clearly demonstrate that graphene junctions with larger connection angles exhibit better thermal-transport properties. In contrast, the effect of connection angle on electronic transport is much more complicated [55], because electrons have more localized distribution and their transport is sensitive to the detailed structure.

3.4.3.5 Graphene quantum dots

Graphene QDs are believed to offer a new approach to quantum nanoelectronics. For considering more realistic models of graphene QDs as proposed in Ref. [53] (Fig. 3.13), we construct graphene QDs by carving narrow constrictions onto a GNR as shown in Fig. 3.18a. Their calculated thermal conductance is presented in Fig. 3.18b. The results show that thermal conductance of graphene QDs is limited by the width of the narrow constriction. Narrowing the connecting region of graphene QDs will strongly suppress thermal conductance of the whole system. The suppression becomes larger

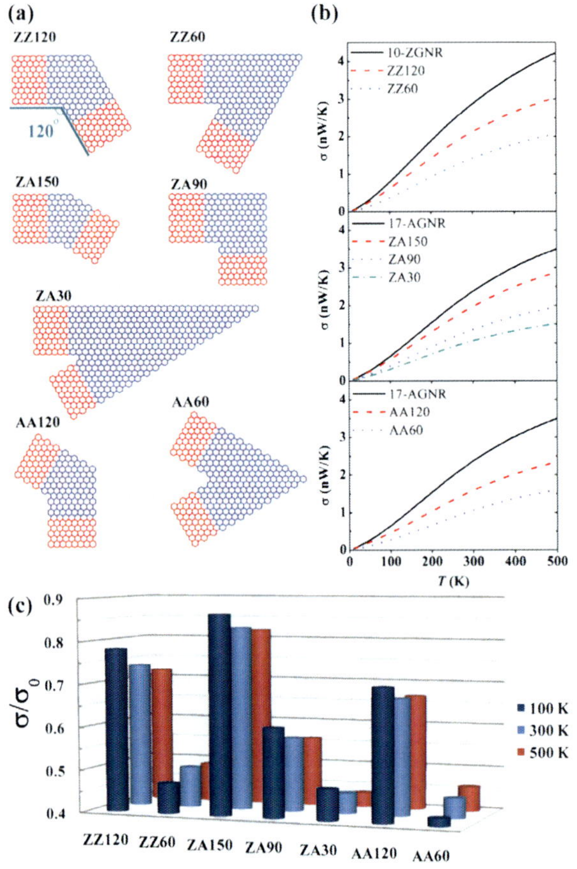

Figure 3.17 (a) Schematics of single-interface junctions formed by connecting 10-ZGNR and 17-AGNR, which have a similar width of around 2.0 nm. The junctions can be classified into three categories in terms of the edge shape and connection angle: (1) "ZZ120" and "ZZ60"; (2) "ZA150", "ZA90" and "ZA30"; and (3) "AA120" and "AA60". (b) Thermal conductance (σ) as a function of temperature. (c) Thermal conductance ratios (σ/σ_0) at 100 K (dark blue), 300 K (blue), and 500 K (red) [51].

at higher temperatures, implying that high-frequency phonons are more affected than the low-frequency phonons. To quantify the magnitude of thermal conductance reduction caused by the QD structure, we calculated the ratio (η) between thermal conductance of the QD and the ballistic thermal conductance of the thermal

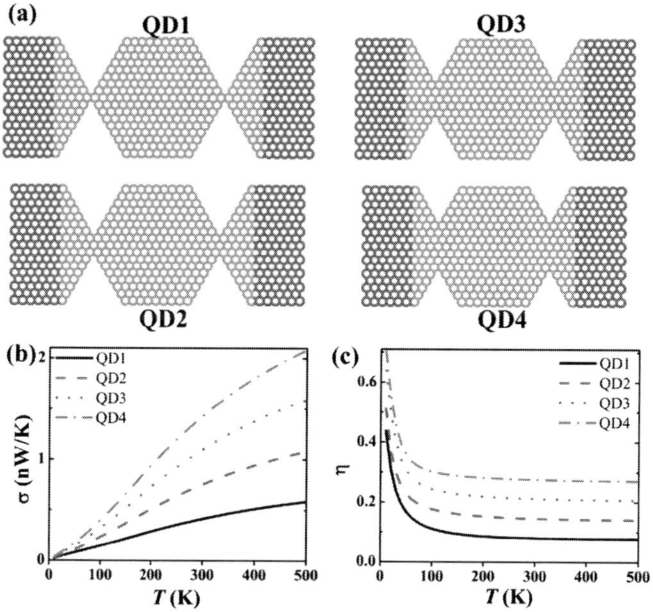

Figure 3.18 (a) Schematics of graphene quantum dots carved from 18-ZGNR. The width of the narrow constrictions increases from "QD1" to "QD4". (b) Thermal conductance (σ) and (c) thermal conductance ratio (η) between quantum dots and the pristine 18-ZGNR as a function of temperature [51].

leads (i.e., pristine GNR). As shown in Fig. 3.18c, η decreases rapidly as the temperature increases, and becomes nearly temperature independent at high temperatures (>100 K for the present systems). The very low η (even lower than 10% at room temperature) indicate that heat dissipation in QDs will be rather slow, which may degrade the performance of the graphene-based QD devices. On the other hand, the extremely low thermal conductance may significantly enhance the thermoelectric efficiency of QD structures.

In short summary, we systematically investigate thermal transport in graphene-based nanodevices, and find that the dependence of thermal-transport properties on the structural characteristics is distinctly different from that of electronic transport. The knowledge and understanding on structure-property relation may be useful for future applications in nanoelectronics and thermoelectricity.

In conclusion, we first demonstrate that the NEGF approach is a very powerful theoretical tool in the study of quantum thermal transport. Then we apply the approach to understand thermal-transport properties of nanostructures, focusing on graphene-related transport systems. The study on thermal transport in GNRs reveals that the extremely high thermal conductivity of graphene is caused by the extraordinary long phonon mean free path. Moreover, the calculations find that the thermal-transport ability of GNRs can be strongly dependent on the edge shape, suggesting to tune thermal conduction by edge control. Furthermore, using graphene-based nanodevices as examples, we apply the NEGF method to explore thermal-transport phenomena in realistic transport systems. Such studies can guide future applications of thermal management and/or thermoelectricity.

References

1. Hone, J., Whitney, M., Piskoti, C., and Zettl, A. (1999) Thermal conductivity of single-walled carbon nanotubes, *Phys. Rev. B*, **59**(4), R2514.
2. Tien, C. L., Majumdar, A., and Gerner, F. M. (1998) *Microscale Energy Transport* (Taylor & Francis, Washington, DC).
3. Kittel, C. (1996) *Introduction to Solid State Physics* (Wiley, New York).
4. Datta, S. (1995) *Electronic Transport in Mesoscopic Systems*, (Cambridge, New York).
5. Rego, L. G. C. and Kirczenow, G. (1998) Quantized thermal conductance of dielectric quantum wires, *Phys. Rev. Lett.*, **81**(1), 232.
6. Schwab, K., Henriksen, E. A., Worlock, J. M., and Roukes, M. L. (2000) Measurement of the quantum of thermal conductance, *Nature*, **404**(6781), 974.
7. Haug, H. and Jauho, A.-P. (1996) *Quantum Kinetics in Transport and Optics of Semiconductors* (Springer, Berlin).
8. Kadanoff, L. P. and Baym, G. (1962) *Quantum Statistical Mechanics*, New York.
9. Wang, J. S., Wang, J., and Lü, J. T. (2008) Quantum thermal transport in nanostructures, *The European Physical Journal B - Condensed Matter and Complex Systems*, **62**(4), 381.

10. Velev, J. and Butler, W. (2004) On the equivalence of different techniques for evaluating the Green function for a semi-infinite system using a localized basis, *J. Phys.: Condens. Matter*, **16**(21), R637.
11. Sancho, M. P. L., Sancho, J. M. L., Sancho, J. M. L., and Rubio, J. (1985) Highly convergent schemes for the calculation of bulk and surface Green functions, *Journal of Physics F: Metal Physics*, **15**(4), 851.
12. Mingo, N. (2006) Anharmonic phonon flow through molecular-sized junctions, *Phys. Rev. B*, **74**(12), 125402.
13. Wang, J.-S., Zeng, N., Wang, J., and Gan, C. K. (2007) Nonequilibrium Green's function method for thermal transport in junctions, *Phys. Rev. E*, **75**(6), 061128.
14. Xu, Y., Wang, J.-S., Duan, W., Gu, B.-L., and Li, B. (2008) Nonequilibrium Green's function method for phonon-phonon interactions and ballistic-diffusive thermal transport, *Phys. Rev. B*, **78**(22), 224303.
15. Geim, A. K. and Novoselov, K. S. (2007) The rise of graphene, *Nat. Mater.*, **6**(3), 183.
16. Novoselov, K. S., Geim, A. K., Morozov, S. V., Jiang, D., Zhang, Y., Dubonos, S. V., Grigorieva, I. V., and Firsov, A. A. (2004) Electric field effect in atomically thin carbon films, *Science*, **306**(5696), 666.
17. Novoselov, K. S., Jiang, D., Schedin, F., Booth, T. J., Khotkevich, V. V., Morozov, S. V., and Geim, A. K. (2005) Two-dimensional atomic crystals, *Proc. Natl. Acad. Sci. U.S.A.*, **102**(30), 10451.
18. Castro Neto, A. H., Guinea, F., Peres, N. M. R., Novoselov, K. S., and Geim, A. K. (2009) The electronic properties of graphene, *Rev. Mod. Phys.*, **81**(1), 109.
19. Li, X., Zhang, G., Bai, X., Sun, X., Wang, X., Wang, E., and Dai, H. (2008). Highly conducting graphene sheets and Langmuir-Blodgett films, *Nature Nanotech.*, **3**(9), 538.
20. Eda, G., Fanchini, G., and Chhowalla, M. (2008) Large-area ultrathin films of reduced graphene oxide as a transparent and flexible electronic material, *Nature Nanotech.*, **3**(5), 270.
21. Bae, S., *et al.* (2010) Roll to-roll production of 30-inch graphene films for transparent electrodes, *Nature Nanotech.*, **5**(8), 574.
22. Kim, K. S., *et al.* (2009) Large-scale pattern growth of graphene films for stretchable transparent electrodes, *Nature*, **457**(7230), 706.
23. Emtsev, K. V., *et al.* (2009) Towards wafer-size graphene layers by atmospheric pressure graphitization of silicon carbide, *Nat. Mater.*, **8**(3), 203.

24. Kim, P., Shi, L., Majumdar, A., and McEuen, P. L. (2001) Thermal transport measurements of individual multiwalled nanotubes, *Phys. Rev. Lett.*, **87**(21), 215502.
25. Pop, E., Mann, D., Wang, Q., Goodson, K., and Dai, H. (2005) Thermal conductance of an individual single-wall carbon nanotube above room temperature, *Nano Lett.*, **6**(1), 96.
26. Li, Q., Liu, C., Wang, X., and Fan, S. (2009) Measuring the thermal conductivity of individual carbon nanotubes by the Raman shift method, *Nanot*, **20**(14), 145702.
27. Hsu, I.-K., Kumar, R., Bushmaker, A., Cronin, S. B., Pettes, M. T., Shi, L., Brintlinger, T., Fuhrer, M. S., and Cumings, J. (2008) Optical measurement of thermal transport in suspended carbon nanotubes, *Appl. Phys. Lett.*, **92**(6), 063119.
28. Balandin, A. A. (2011) Thermal properties of graphene and nanostructured carbon materials, *Nat. Mater.*, **10**(8), 569.
29. Balandin, A. A., Ghosh, S., Bao, W., Calizo, I., Teweldebrhan, D., Miao, F., and Lau, C. N. (2008) Superior thermal conductivity of single-layer graphene, *Nano Lett.*, **8**(3), 902.
30. Ghosh, S., Calizo, I., Teweldebrhan, D., Pokatilov, E. P., Nika, D. L., Balandin, A. A., Bao, W., Miao, F., and Lau, C. N. (2008) Extremely high thermal conductivity of graphene: Prospects for thermal management applications in nanoelectronic circuits, *Appl. Phys. Lett.*, **92**(15), 151911.
31. Cai, W., Moore, A. L., Zhu, Y., Li, X., Chen, S., Shi, L., and Ruoff, R. S. (2010) Thermal transport in suspended and supported monolayer graphene grown by chemical vapor deposition, *Nano Lett.*, **10**(5), 1645.
32. Faugeras, C., Faugeras, B., Orlita, M., Potemski, M., Nair, R. R., and Geim, A. K. (2010) Thermal conductivity of graphene in corbino membrane geometry, *ACS Nano*, **4**(4), 1889.
33. Jauregui, L. A., *et al.* (2010) Thermal transport in graphene nanostructures: Experiments and simulations, *ECS Transactions*, **28**(5), 73.
34. Seol, J. H., *et al.* (2010) Two-dimensional phonon transport in supported graphene, *Science*, **328**(5975), 213.
35. Mohanty, N., Moore, D., Xu, Z., Sreeprasad, T. S., Nagaraja, A., Rodriguez, A. A., and Berry, V. (2012) Nanotomy-based production of transferable and dispersible graphene nanostructures of controlled shape and size, *Nat. Commun.*, **3**(1), 844.
36. Kosynkin, D. V., Higginbotham, A. L., Sinitskii, A., Lomeda, J. R., Dimiev, A., Price, B. K., and Tour, J. M. (2009) Longitudinal unzipping of carbon nanotubes to form graphene nanoribbons, *Nature*, **458**(7240), 872.

37. Jiao, L., Zhang, L., Wang, X., Diankov, G., and Dai, H. (2009) Narrow graphene nanoribbons from carbon nanotubes, *Nature*, **458**(7240), 877.

38. Xu, Y., Chen, X., Gu, B.-L., and Duan, W. (2009) Intrinsic anisotropy of thermal conductance in graphene nanoribbons, *Appl. Phys. Lett.*, **95**(23), 233116.

39. Brenner, D. W., Shenderova, O. A., Harrison, J. A., Stuart, S. J., Ni, B., and Sinnott, S. B. (2002) A second-generation reactive empirical bond order (REBO) potential energy expression for hydrocarbons, *J. Phys.: Condens. Matter*, **14**(4), 783.

40. Vandescuren, M., Hermet, P., Meunier, V., Henrard, L., and Lambin, P. (2008) Theoretical study of the vibrational edge modes in graphene nanoribbons, *Phys. Rev. B*, **78**(19), 195401.

41. Mingo, N. and Broido, D. A. (2005) Carbon nanotube ballistic thermal conductance and its limits, *Phys. Rev. Lett.*, **95**(9), 096105.

42. Saito, K., Nakamura, J., and Natori, A. (2007) Ballistic thermal conductance of a graphene sheet, *Phys. Rev. B*, **76**(11), 115409.

43. Markussen, T., Jauho, A.-P., and Brandbyge, M. (2008) Heat conductance is strongly anisotropic for pristine silicon nanowires, *Nano Lett.*, **8**(11), 3771.

44. Xu, Y., Zou, X., Chen, X., and Duan, W. Thermoelectric properties of boron nitride nanotubes and nanoribbons, in preparation.

45. Gonze, X., *et al.* (2009) ABINIT: First-principles approach to material and nanosystem properties, *Comput. Phys. Commun.*, **180**(12), 2582.

46. Zabel, H. (2001) Phonons in layered compounds, *J. Phys.: Condens. Matter*, **13**(34), 7679.

47. Jiang, J.-W., Wang, J.-S., and Li, B. (2009) Thermal conductance of graphene and dimerite, *Phys. Rev. B*, **79**(20), 205418.

48. Zhu, H., Xu, Y., Gu, B.-L., and Duan, W. (2012) Robust linear dependence of thermal conductance on radial strain in carbon nanotubes, *New J. Phys.*, **14**(1), 013053.

49. Mingo, N., Stewart, D. A., Broido, D. A., and Srivastava, D. (2008) Phonon transmission through defects in carbon nanotubes from first principles, *Phys. Rev. B*, **77**(3), 033418.

50. Huang, H., Xu, Y., Zou, X., Wu J., and Duan, W. Tuning thermal conduction via extended defects in graphene, *Phys. Rev. B*, **87**(20), 205415.

51. Xu, Y., Chen, X., Wang, J.-S., Gu, B.-L., and Duan, W. (2010) Thermal transport in graphene junctions and quantum dots, *Phys. Rev. B*, **81**(19), 195425.

52. Yan, Q., Huang, B., Yu, J., Zheng, F., Zang, J., Wu, J., Gu, B.-L., Liu, F., and Duan, W. (2007) Intrinsic current-voltage characteristics of graphene nanoribbon transistors and effect of edge doping, *Nano Lett.*, **7**(6), 1469.
53. Westervelt, R. M. (2008) Graphene nanoelectronics, *Science*, **320**(5874), 324.
54. Ponomarenko, L. A., Schedin, F., Katsnelson, M. I., Yang, R., Hill, E. W., Novoselov, K. S., and Geim, A. K. (2008) Chaotic dirac billiard in graphene quantum dots, *Science*, **320**(5874), 356.
55. Areshkin, D. A. and White, C. T. (2007) Building blocks for integrated graphene circuits, *Nano Lett.*, **7**(11), 3253.
56. Li, X., Wang, X., Zhang, L., Lee, S., and Dai, H. (2008) Chemically derived, ultrasmooth graphene nanoribbon semiconductors, *Science*, **319**(5867), 1229.
57. Swartz, E. T. and Pohl, R. O. (1989) Thermal boundary resistance, *Rev. Mod. Phys.*, **61**(3), 605.
58. Chen, Z., Jang, W., Bao, W., Lau, C. N., and Dames, C. (2009) Thermal contact resistance between graphene and silicon dioxide, *Appl. Phys. Lett.*, **95**(16), 161910.
59. Campos, L. C., Manfrinato, V. R., Sanchez-Yamagishi, J. D., Kong, J., and Jarillo-Herrero, P. (2009) Anisotropic eching and nanoribbon formation in single-layer graphene, *Nano Lett.*, **9**(7), 2600.

Chapter 4

Ballistic Thermal Transport by Phonons at Low Temperatures in Low-Dimensional Quantum Structures

Zhong-Xiang Xie and Ke-Qiu Chen
Department of Applied Physics, Hunan University, Changsha 410082, China
keqiuchen@hnu.edu.cn

Ballistic thermal transport by phonons at low temperature is analyzed in low-dimensional quantum structures modulated with inhomogeneities. First, within the Landauer transport theory, we present a general formula to calculate the ballistic thermal conductance associated with phonons in the linear response limit. Then, we introduce a continuum elastic model to describe the elastic phonon waves with long wavelength and provide the indispensable details of the scattering-matrix method to calculate the transmission coefficient of elastic waves in the low-dimensional structures. According to the numerical results, the effects of boundary conditions, dimensions, and inhomogeneities (including abrupt junctions, structural defects, etc.) on the ballistic thermal transport by phonons are discussed in detail. A comparative analysis for the ballistic thermal transport is made between two-dimensional (2D) and three-dimensional (3D) models. Some interesting physical

Nanoscale Energy Transport and Harvesting: A Computational Study
Edited by Gang Zhang

ISBN 978-981-4463-02-7 (Hardcover), 978-981-4463-03-4 (eBook)
www.panstanford.com

properties are revealed, such as the universal quantum thermal conductance, periodic transmission behaviors, stop-frequency gap, and so on. These results will be helpful to understand the behaviors of ballistic phonon thermal transport in quantum structures with inhomogeneities, and to the design of thermal quantum devices.

4.1 Introduction

In the past two decades, the electronic transport properties in nanoscale structures have attracted considerable attention [1–3], since they possess novel physical properties in comparison with bulk materials, as well as potential applications in new devices. Now it has become well known that the electronic properties can be engineered to a high degree of precision in quantum structures by using quantum size effects on electrons. On the other hand, currently there is an increasing interest in the thermal properties due to practical needs and fundamental science. As known, heat needs to be removed as much as possible, and the search for specific materials or structures that conduct heat well has become essential for the design of new integrated circuits and 3D electronics [4], due to the fact that a continuous scaling down of the sizes of these devices leads to an increase in power dissipation per unit area. However, for the thermoelectric energy conversion, it needs materials or structures of a strongly suppressed thermal conductance in order to improve their thermoelectric efficiency [5–7]. Especially, for the functional thermal devices, such as thermal rectifier [8–9], thermal transistor [10], thermal diode [11], thermal memory [12], and so on, the heat fluxes must be reasonably tunable to implement special given thermal functions. Therefore, scientifically exploring the thermal-transport mechanisms in quantum systems becomes critically important from a physics perspective and for practical applications of various nanodevices.

In recent years, the thermal transport associated with phonons in low-dimensional nanostructures has attracted increasing attention both experimentally [13–19] and theoretically [20–27]. It has been demonstrated that the thermal conductivity of nanostructures is much smaller than their bulk values due to the boundary confinement. In very low temperatures, due to the fact that thermal wavelength becomes comparable to or larger than the dimensions of the thermal pathway, the phonons transport ballistically. Following the Landauer transport theory originally used for the electrical conductance [28, 29], several groups have derived the expressions of thermal conductance by ballistic phonons in low temperature regions [20–23], and predicted that the thermal conductance at low temperature is quantized in a universal unit $\pi^2 k_B^2 T/3h$ [20, 24], similar to the role of $2e^2/h$ as a quantum of electrical conductance in one-dimensional wires. Afterwards, these predictions were verified in a beautiful experiment by Schwab *et al.* [14]. Their experiment shows a result very consistent with the quantum value at temperatures below about 0.08 K. To understand the quantum thermal conductance behaviors, many efforts have been devoted to elucidate the thermal-transport mechanisms in various nanostructures, such as thin plates [25], nanowires [30–32], nanoribbons [33–38], superlattices [39–41], nanotubes [42–46], and so on. Now, it has become well known that unlike the diffusive phonon transport at high temperatures in bulk materials, the thermal current in the ballistic region is mainly carried by a set of the discrete vibrational phonon modes due to the phonon boundary confinement in the quantum systems. Therefore, in the low temperature limit, each mode will contribute $K_0(K_0 = \pi^2 k_B^2 T/3h)$ to the total thermal conductance; a perfect sample has $N_A K_0$ with N_A being the number of acoustic modes with zero cutoff frequency at long wavelengths.

However, the smooth quantum plateau of the low temperature thermal conductance is only presented if the transmission or thermal phonons in a nanowire is perfect. For most cases, it is unexpected that the quantum platform is prematurely destroyed and instead the thermal conductance is decreased below the quantum value as the temperature goes up in the low temperature range. It is suggested that the decreases of the ballistic thermal conductance in low temperature range may stem from the interface scattering due to inhomogeneities such as abrupt junctions between the wire and the reservoirs [47], rough surfaces [48–50], structural defects [51–53] in the quantum structure, and so on. Based on the full elasticity theory, Santamore and Cross investigated the effect of surface roughness on the ballistic thermal conductance in a dielectric nanowire in low temperature range [48]. They showed that the presence of rough surface significantly decreases the thermal conductance below the quantum value. Chen *et al.* studied the influence of defects on the low temperature thermal conductance contributed from several lowest horizontally polarized shear (SH) modes in the nanowire by using the scattering-matrix method, and found an obvious decrease of the lattice thermal conductance in the defective nanowires [51]. More recently, Peng *et al.* calculated the thermal conductance associated with the lowest six types ballistic phonon modes in quantum wire with catenoidal contacts [54], and observed the destruction of the quantum thermal conductance plateau and a decrease of thermal conductance appeared in such a structure when the temperature is sufficiently low. In addition, the decrease of the thermal conductance was also analyzed in other quantum waveguide, such as stub structures [55, 56], T-shaped quantum wires [57, 58], double-bend nanowires [59, 60], and so on. From these results, one knows that these inhomogeneities actually act as the thermal resistance, which leads to decrease the phonon thermal conductance, and such thermal resistance sensitively depends on the intrinsic properties and sizes of configurations [55–61].

In spite of all these advancements mentioned above, a further understanding of the ballistic thermal transport by phonons remains to be important for both foundational significance and applications in many nanodevices. Recently, Wang *et al.* presented a review of the methods for thermal-transport calculations for nanojunctions connected to two heat reservoirs [62], in which their emphasis is on fundamental quantum theory and atomistic models. Based on the single-particle scattering approach, non-equilibrium Green's function formalism, molecular dynamics, etc., Dubi *et al.* systematically review the thermal conductance, local temperature and heating, and thermoelectricity in nanoscale systems, such as molecular junctions, suspended nanotubes, quantum point contacts, and so on [63]. Similar to electrons, it is proved that phonons can be put to beneficial use, such as to carry and process information. In a recent colloquium, Li *et al.* discuss in detail the manipulation of thermal flow on the nanoscale and the task of processing information by utilizing phonons, and particularly focus on the basic, elementary building blocks for handling phonons, namely the conceptual realization and its possible operation of some function thermal (phonon) devices mentioned above [64]. According to our previous works, here we present a review of the low temperature ballistic thermal transport contributed by phonons in quantum structures, of particular focus on the effects of the boundary conditions, dimensions, and inhomogeneities on the ballistic thermal conductance. In Section 4.2, we introduce the Landauer formalism for calculating the ballistic thermal conductance in nanoscale systems and present the necessary details of the scattering-matrix method within the framework of elastic continuum model. In Section 4.3, we discuss numerically the ballistic thermal-transport properties of the elastic waves in both 2D and 3D models. In Section 4.4, we provide a summary of the study.

4.2 Formalism

This section is divided into three parts. First we embark on the Landauer formula for the phonon thermal conductance in the linear response limit, analogous to the electrical conductance formula. Then we present a continuum elastic model to describe the ballistic phonon transport. Finally, we give the scattering-matrix method of calculating the phonon transmission coefficient for quite a general geometric structure shown in Fig. 4.1.

4.2.1 *Landauer Formula for the Thermal Conductance*

We assume that a two-terminal system is connected to hot and cold reservoirs with a small temperature difference δT. Here $\delta T = T_{\text{hot}} - T_{\text{cold}}$ and $\delta T \ll T_{\text{hot}}(T_{\text{cold}})$, that T_{hot} and T_{cold} denote the temperature of the hot and cold reservoirs, respectively. So we can adopt the mean temperature $T = (T_{\text{hot}} + T_{\text{cold}})/2$ as the temperature of the hot and cold reservoirs. According to the Landauer theory [20–22, 25], the heat flux J from the hot reservoir to the cold reservoir can be expressed by

$$J = \sum_m \int_0^\infty \frac{dk}{2\pi} \hbar\omega_m(k) v_m(k) (f_{\text{hot}} - f_{\text{cold}}) \tau_m(k). \qquad (4.1)$$

Here m denotes the index of vibrational modes, $f = 1/\left[\exp(\hbar\omega/k_B T) - 1\right]$ is the thermal distribution of phonons in the hot (cold) reservoir. $\tau_m(k)$ is the phonon transmission coefficient through all the system, $\omega_m(k)$ and $v_m(k)$ are the frequency and group velocity of the mode m. Since the group velocity can be canceled by $v_m = \partial\omega_m/\partial k$, Eq. 4.1 becomes

$$J = \sum_m \int_0^\infty \frac{d\omega}{2\pi}\hbar\omega(f_{\text{hot}} - f_{\text{cold}})\tau_m(\omega). \tag{4.2}$$

Note that within this approach the electronic contribution to the heat flux is calculated in a similar way as the phonon case with the substitutions $\hbar\omega \to E - \mu$ and $f_{\text{hot(cold)}} \to g_{\text{hot(cold)}}$, where E and μ are the electronic energy and the chemical potential, $g_{\text{hot(cold)}}$ is the Fermi–Dirac distribution function, which is also temperature dependent. The thermal conductance is given by $K = J/\delta T$ in the limit $\delta T = T_{\text{hot}} - T_{\text{cold}} \to 0$. Therefore, at $T = (T_{\text{hot}} + T_{\text{cold}})/2$ we have

$$\begin{aligned} K &= \frac{1}{2\pi}\sum_m \int_{\omega_{\text{m}}}^\infty d\omega\hbar\omega\frac{\partial f(\omega, T)}{\partial T}\tau_m(\omega) \\ &= \frac{\hbar^2}{2\pi k_B T^2}\sum_m \int_{\omega_{\text{m}}}^\infty d\omega\hbar\omega\frac{\omega^2 e^{\hbar\omega/k_B T}}{(e^{\hbar\omega/k_B T} - 1)^2}\tau_m(\omega), \end{aligned} \tag{4.3}$$

with ω_{m} denoting the cutoff frequency of the mode m. Introducing the scaled frequency variable $x = \hbar\omega/k_B T$ gives the expression as below

$$K = \frac{k_B^2 T}{h}\sum_m \int_{x_m}^\infty dx\frac{x^2 e^x}{(e^x - 1)^2}\tau_m\left(\frac{k_B T}{\hbar}x\right). \tag{4.4}$$

For the moment, let us assumed that the adiabatic contact between the two thermal reservoirs and the quantum structure is perfect, so that $\tau_m = 1$. For this idealized case, Eq. 4.4 shows that the quantity $k_B^2 T/h$ plays the role of the quantum unit of the thermal conductance, similar to the role of e^2/h as the quantum electrical conductance for one-dimensional wires. In the limit $T \to 0$, the contribution of the thermal conductance stems from several acoustic modes with nonzero cutoff frequency, i.e., $x_m = 0$. After the integration of Eq. 4.4, we have $K = N_A K_0$, where N_A denotes the number of acoustic modes with zero cutoff frequency.

However, for most of the realistic cases, the contact between the thermal reservoirs and the quantum structure is generally imperfect or the incoming phonons are always scattered due to inhomogeneities. At this point, $\tau_m(\omega) < 1$. Therefore, a central issue in calculating the thermal conductance determined by Eq. 4.4 is then to obtain the transmission coefficient $\tau_m(\omega)$. To calculate $\tau_m(\omega)$, several methods such as scattering or transfer matrices method [54–61, 65, 66], boundary condition method [67], mode-matching method [68], non-equilibrium Green's function method [42, 43, 62], have been developed to obtain the transmission coefficient. All these methods are fundamentally equivalent, and in fact stem from the single–particle elastic scattering theory of conduction. Among these methods, the transmission coefficient is generally written as a sum of all the transmission coefficient $\tau_{mn}(\omega)$, where $\tau_{mn}(\omega)$ denotes the transmission coefficient from the m phonon mode in the incoming lead at frequency ω to the n phonon mode in the outgoing lead, namely

$$\tau(\omega) = \sum_m \sum_n \tau_{mn}(\omega) = \mathrm{Trace}(\sigma\sigma^\dagger), \tag{4.5}$$

where σ is a $N_\mathrm{O} \times N_\mathrm{I}$ scattering matrix. Here, N_I and N_O represent the number of phonon channels at frequency ω that can be opened in the incoming and outgoing leads, respectively. However, in real calculations, we should use a sufficiently large σ to consider the contributions from evanescent modes. Based on the non-equilibrium Green's functions, the transmission coefficient can be also cast in anther equivalent form

$$\tau(\omega) = \mathrm{Trace}(G^r\Gamma_L G^a \Gamma_R), \tag{4.6}$$

where G^r ($G^a = (G^r)^\dagger$) is the retarded (advanced) Green's function for the central scattering region and $\Gamma_{L(R)}$ represents the coupling interaction between the central scattering region and the left (right) electrode. This result is first derived for the electronic transport. Recently, it has been repeatedly derived for thermal transport based on atomistic models [69–73], and widely applied in exploring the thermal properties in real materials [74, 75]. Here, we mainly introduce the scattering-matrix method to calculate the transmission coefficient $\tau_m(\omega)$ of acoustic phonons within the continuum elastic model, which will be discussed later.

4.2.2 Continuum Elastic Model

In the low temperature region, the thermal conductance is mainly attributed to the contributions of several lowest phonon modes with low frequency. The wavelength of these phonons is generally much larger than any microscopic length, such as the atomic bond length. Thus, the continuum elastic model is quite adaptive to describe the ballistic phonon transport in quantum structures. This is also confirmed by microscopic calculations [76] and Raman experiments [77] for dielectric continuum model and by phonon transmission spectroscopy [78] for the continuum elastic model. According to continuum elastic theory, the phonon displacement field $\mathbf{U}(u_x, u_y, u_z)$ in the Cartesian coordinates can be given in terms of the scalar potential ϕ and the vector potential $\mathbf{H}$ [79–81]

$$\mathbf{U} = \nabla\phi + \nabla \times \mathbf{H} \tag{4.7}$$

with $\mathbf{H}= (H_x, H_y, H_z)$ and $\nabla = \partial/\partial x\vec{i} + \partial/\partial y\vec{j} + \partial/\partial z\vec{k}$. Each potential function satisfies the following wave equations:

$$v_L^2\nabla^2\phi + \omega^2\phi = 0, \tag{4.8}$$

$$v_T^2\nabla^2 H_x + \omega^2 H_x = 0, \tag{4.9}$$

$$v_T^2\nabla^2 H_y + \omega^2 H_y = 0, \tag{4.10}$$

$$v_T^2\nabla^2 H_z + \omega^2 H_z = 0, \tag{4.11}$$

with $\nabla^2 = \partial^2/\partial x^2 + \partial^2/\partial y^2 + \partial^2/\partial z^2$. Here, ω is the frequency of the wave and v_L, v_T denote the sound velocities of the bulk longitudinal and transverse acoustic waves, respectively, and given as

$$v_L = \sqrt{\frac{\lambda + 2\mu}{\rho}} \tag{4.12}$$

$$v_T = \sqrt{\frac{\mu}{\rho}}, \tag{4.13}$$

where ρ and $\lambda(\mu)$ represent the mass density and the Lame coefficients, respectively. According to Eq. 4.7, the displacement field $\mathbf{U}$ can be divided in to three component forms as

$$u_x = \frac{\partial\phi}{\partial x} + \frac{\partial H_z}{\partial y} - \frac{\partial H_y}{\partial z}, \tag{4.14}$$

$$u_y = \frac{\partial \phi}{\partial y} - \frac{\partial H_z}{\partial x} + \frac{\partial H_x}{\partial z}, \tag{4.15}$$

$$u_z = \frac{\partial \phi}{\partial z} + \frac{\partial H_y}{\partial x} - \frac{\partial H_x}{\partial y}. \tag{4.16}$$

According to three components above, the corresponding stress can be derived as

$$T_{xx} = \lambda \left(\frac{\partial u_x}{\partial x} + \frac{\partial u_y}{\partial y} + \frac{\partial u_z}{\partial z} \right) + 2\mu \frac{\partial u_x}{\partial x}, \tag{4.17}$$

$$T_{yy} = \lambda \left(\frac{\partial u_x}{\partial x} + \frac{\partial u_y}{\partial y} + \frac{\partial u_z}{\partial z} \right) + 2\mu \frac{\partial u_y}{\partial y}, \tag{4.18}$$

$$T_{zz} = \lambda \left(\frac{\partial u_x}{\partial x} + \frac{\partial u_y}{\partial y} + \frac{\partial u_z}{\partial z} \right) + 2\mu \frac{\partial u_z}{\partial z}, \tag{4.19}$$

$$T_{xy} = T_{yx} = \mu \left(\frac{\partial u_x}{\partial y} + \frac{\partial u_y}{\partial x} \right), \tag{4.20}$$

$$T_{yz} = T_{zy} = \mu \left(\frac{\partial u_y}{\partial z} + \frac{\partial u_z}{\partial y} \right), \tag{4.21}$$

$$T_{xz} = T_{zx} = \mu \left(\frac{\partial u_z}{\partial x} + \frac{\partial u_x}{\partial z} \right). \tag{4.22}$$

For a 2D system, at hand the displacement vector **U** is independent of the z coordinate and one can take $\partial/\partial z = \partial^2/\partial z^2 = 0$, Eqs. 4.14–4.16 can be written as

$$u_x = \frac{\partial \phi}{\partial x} + \frac{\partial H_z}{\partial y}, \tag{4.23}$$

$$u_y = \frac{\partial \phi}{\partial y} - \frac{\partial H_z}{\partial x}, \tag{4.24}$$

$$u_z = \frac{\partial H_y}{\partial x} - \frac{\partial H_x}{\partial y}. \tag{4.25}$$

From Eqs. 4.23–4.25, one knows that a 2D system can support two independent modes: the in-plane modes (u_x, u_y) and the out-of-plane mode (u_z) (determined by Eqs. 4.23 and 4.24, and by Eq. 4.25), respectively. Thus, it is convenient that they can be done independently to avoid dealing with the vector wave equations. According to the normalization condition of the vector potential $\Delta\mathbf{H} = 0$, H_x, H_y can be easily eliminated and then Eq. 4.25 can be further simplified as

$$v_T^2\nabla^2 u_z+\omega^2 u_z=0. \tag{4.26}$$

Interestingly, Eq. 4.26, which describes the vibrational displacement of the out-of-mode, has formally the same structure as Eqs. 4.9–4.11. As expounded in the elastic textbook by Graff [81], the in-plane modes are generally split into the longitudinal polarized P wave, the vertically polarized SV wave, and the out-of-plane mode is renamed as the horizontally polarized shear SH wave. Their polarization directions are along the x, y, and z directions, respectively. Note that because both the P wave and the SV wave in x–y plane couple with each other, the detailed discussion of their ballistic transmission properties in quantum systems becomes very complicated, which is also different from the case for the SH wave.

4.2.3 *Scattering-Matrix Method*

For the convenience of elaboration in the following text, we present a simple 2D model structure, which has been done in many previous studies, as shown in Fig. 4.1. In such a structure, we usually divide all the system into three regions: the left semi-infinite region I, the central region II, and the right semi-infinite region III, of which the corresponding transverse width is denoted by W_I, W_{III} and W_{II}, respectively. It is assumed that the contacts between the two leads and reservoirs are perfect without any reflection and the elastic scattering only occurs at attached interfaces between two blocks with different width.

Firstly, we consider the simplest case for the SH wave incidence, which is described by

$$v_T^2\nabla^2\psi+\omega^2\psi^2=0 \tag{4.27}$$

with $\psi=u_z$, $\nabla^2 = \partial^2/\partial x^2 + \partial^2/\partial y^2$. The solution of the above equation defined in each region ξ (ξ = I, II, and III) are given in terms of linear combination of the transmitted and reflected waves

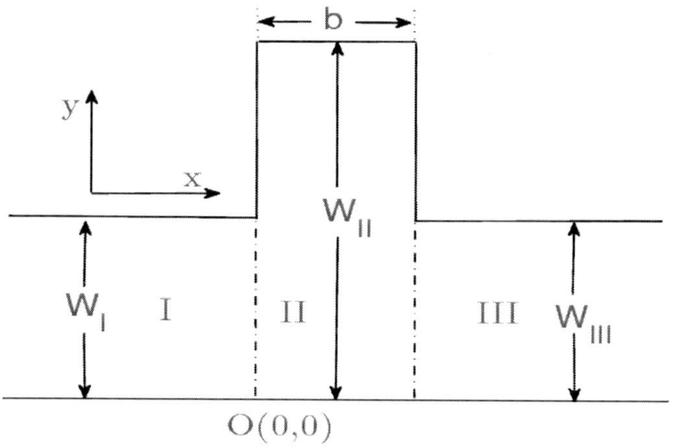

Figure 4.1 Schematic diagram the T-shaped quantum structure.

propagating along the x direction as follows:

$$\psi^{\xi}(x,y) = \sum_{n=0}^{N^{\xi}} [A_n^{\xi} e^{ik_n^{\xi}(x-x_{\xi})} + B_n^{\xi} e^{-ik_n^{\xi}(x-x_{\xi})}]\eta_n^{\xi}(y), \tag{4.28}$$

where x_{ξ} is the reference coordinate in region ξ along the x axis, $\eta_n^{\xi}(y)$ represents the orthonormal transverse wave function of mode n in region ξ, A_n^{ξ}, and B_n^{ξ} are constants to be determined by matching the boundary conditions. When the stress-free boundary condition ($\hat{n} \cdot \nabla\psi = 0$) is employed at the edges with $\hat{n}$ being the vector perpendicular to the edge, the transverse function $\eta_n^{\xi}(y)$ can be expressed by

$$\eta_n^{\xi}(y) = \begin{cases} \sqrt{2/W_{\xi}} \cos(n\pi y/W_{\xi}), & \text{if } n \neq 0 \\ \sqrt{1/W_{\xi}}, & \text{if } n = 0 \end{cases}. \tag{4.29}$$

However, when the hard wall boundary condition ($\psi = 0$) is employed at the edges, it should be expressed as

$$\eta_n^{\xi}(y) = \sqrt{2/W_{\xi}} \sin(n\pi y/W_{\xi}), \quad \text{if } n \neq 0, \tag{4.30}$$

which is similar to the case of the electron propagation. It is worth noting that the stress-free boundary condition of the acoustic phonon allows the propagation of an mode with $\omega = 0$ as one expect for elastic waves, while the hard wall boundary condition does not. Because of many similarities in the derived process for these two conditions, here we only show a case for the stress-free boundary condition to reduce the length of the review. In Eq. 4.28, k_n^ξ is the corresponding wave vector along the x axis, given by the frequency matching condition

$$k_n^\xi = \sqrt{\frac{\omega^2 - (\omega_n^\xi)^2}{v_T^2}}, \tag{4.31}$$

where $\omega_n^\xi = n\pi v_T / W_\xi$ is the cutoff frequency of the mode n in the region ξ. As known, for the "open" channels, k_n^ξ must be real, i.e., those with $\omega > \omega_n^\xi$. Otherwise, we take $k_n^\xi = i|k_n^\xi|$.

Note that although only the propagating modes will contribute to the wave transmission, the sums over n in Eq. 4.28 include all the propagating and evanescent modes (imaginary k). However, in real calculations, one generally takes all the propagating modes and several lowest evanescent modes into account to meet the desired precision. This is also confirmed by these previous studies [65, 82]. For the interface ($x=0$) between regions I and II, the displacement ψ and stress should be continuous in terms of the boundary matching condition, i.e.,

$$\psi^I(x = 0, y) = \psi^{II}(x = 0, y), \tag{4.32}$$

$$\mu[\partial\psi^I(x, y)/\partial x]|_{x=0} = \mu[\partial\psi^{II}(x, y)/\partial x]|_{x=0}. \tag{4.33}$$

Now substituting Eq. 4.28 into Eqs. 4.32 and 4.33, we can easily get

$$\sum_{n=0}^{N^I} (A_n^I + B_n^I)\eta_n^I(y) = \sum_{n=0}^{N^{II}} (A_n^{II} + B_n^{II})\eta_n^{II}(y), \tag{4.34}$$

$$\sum_{n=0}^{N^I} k_n^I (A_n^I - B_n^I)\eta_n^I(y) = \sum_{n=0}^{N^{II}} k_n^{II} (A_n^{II} - B_n^{II})\eta_n^{II}(y). \tag{4.35}$$

To extract the relationships of four constants in these equations, one multiplies two sides of Eq. 4.34 with $\eta_m^I(y)$ and two sides of Eq. 4.35 with $\eta_m^{II}(y)$ and integrates over the transverse width of the central region. By using the orthogonality of the wave functions over the appropriate domain, i.e., $\int_0^{W_\xi} \eta_m^\xi(y)\eta_n^\xi(y)dy = \delta_{mn}$, we get

$$A_m^I + B_m^I = \sum_{n=0}^{N^{II}} D_{mn}(A_n^{II} + B_n^{II}), \tag{4.36}$$

$$\sum_{n=0}^{N^I} D_{nm} k_m^I (A_m^I - B_m^I) = k_n^{II} (A_n^{II} - B_n^{II}). \tag{4.37}$$

This is a set of linear equations that determine the relationships between these expansion coefficients. In these above equations, D_{mn} represents the coupling strength of different modes, which is expressed by the overlap of the wave functions for the regions I and II

$$D_{mn} = \int_0^{W_I} \eta_m^I(y) \eta_n^{II}(y) dy. \tag{4.38}$$

Substituting Eq. 4.29 into the above equation, we can derive the explicit form of D_{mn}, i.e.,

$$D_{mn} = \sqrt{\frac{W_I}{W_{II}}}, \quad \text{if } m=0, n=0; \tag{4.39}$$

$$D_{mn} = \frac{1}{n\pi} \sqrt{\frac{2W_{II}}{W_I}} \sin\left(\frac{n\pi W_I}{W_{II}}\right), \quad \text{if } m = 0,\ n \neq 0; \tag{4.40}$$

$$D_{mn} = 0, \quad \text{if } m \neq 0,\ n = 0; \tag{4.41}$$

$$D_{mn} = \frac{1}{\sqrt{W_I W_{II}}} \left[\frac{1}{\frac{m\pi}{W_I} + \frac{n\pi}{W_{II}}} \sin\left(m\pi + \frac{n\pi W_I}{W_{II}}\right) + \frac{1}{\frac{m\pi}{W_I} - \frac{n\pi}{W_{II}}} \sin\left(m\pi - \frac{n\pi W_I}{W_{II}}\right)\right], \text{ if } m \neq 0,\ n \neq 0,\ \frac{m}{W_I} \neq \frac{n}{W_{II}}; \tag{4.42}$$

$$D_{mn} = \sqrt{\frac{W_I}{W_{II}}}, \quad \text{if } m \neq 0,\ n \neq 0,\ \frac{m}{W_I} = \frac{n}{W_{II}}. \tag{4.43}$$

For the sake of making the matrix calculation, one defines matrices D, K^I, K^{II} whose elements are D_{mn}, $k_m^I \delta_{mn}$, $k_n^{II} \delta_{mn}$, respectively. Similarly, four column vectors A^I, B^I, A^{II}, and B^{II} are also constructed by the corresponding elements A_m^I, B_m^I, A_n^{II}, and B_n^{II}. In this notation, we have

$$A^I + B^I = D(A^{II} + B^{II}), \tag{4.44}$$

$$D^T K^I (A^I - B^I) = K^{II}(A^{II} - B^{II}). \tag{4.45}$$

Note that Eqs. 4.44 and 4.45 are series of homogeneous linear equations, and can be rewritten in the form of matrix

$$\begin{pmatrix} A^I \\ B^I \end{pmatrix} = M(I, II) \begin{pmatrix} A^{II} \\ B^{II} \end{pmatrix}, \tag{4.46}$$

where

$$M(I, II) = \begin{pmatrix} M_{11}(I, II) & M_{12}(I, II) \\ M_{21}(I, II) & M_{22}(I, II) \end{pmatrix}$$
$$= \begin{pmatrix} E^I & E^I \\ D^T K^I & -D^T K^I \end{pmatrix}^{-1} \begin{pmatrix} D & D \\ K^{II} & -K^{II} \end{pmatrix} \tag{4.47}$$

with E^I being the identity matrix. In fact, $M(I, II)$ is the resulting transfer matrix. However, it is well known that the numerical scheme with the transfer matrix in general becomes unstable, since there waves which vary exponentially with the distance [83, 84], especially at very high frequencies. In order to avoid this numerical instability, one introduces the scattering matrix

$$S(I, II) = \begin{pmatrix} S_{11}(I, II) & S_{12}(I, II) \\ S_{21}(I, II) & S_{22}(I, II) \end{pmatrix} \tag{4.48}$$

at the interface ($x = 0$) between regions I and II, which is defined by

$$\begin{pmatrix} A^{II} \\ B^I \end{pmatrix} = \begin{pmatrix} S_{11}(I, II) & S_{12}(I, II) \\ S_{21}(I, II) & S_{22}(I, II) \end{pmatrix} \begin{pmatrix} A^I \\ B^{II} \end{pmatrix}. \tag{4.49}$$

With the help of the transfer matrix $M(I, II)$, the elements of $S(I, II)$ are given by

$$\begin{aligned} S_{11}(I, II) &= M_{11}(I, II) - M_{12}(I, II)(M_{22}(I, II))^{-1} M_{21}(I, II) \\ S_{12}(I, II) &= M_{12}(I, II)(M_{22}(I, II))^{-1} \\ S_{21}(I, II) &= (M_{22}(I, II))^{-1} M_{21}(I, II) \\ S_{22}(I, II) &= (M_{22}(I, II))^{-1}. \end{aligned} \tag{4.50}$$

As shown by Li *et al.* [85], the scattering matrix connecting the left and right of the region II can be written as

$$S(II, II) = \begin{pmatrix} 0 & P \\ P & 0 \end{pmatrix}, \tag{4.51}$$

where P is an N-dimensional diagonal matrix with diagonal elements $P_{mn} = e^{ik_n^{II} b} \delta_{mn}$, and b is the longitudinal length of region II.

By using the similar scheme, the scattering matrix $S(II, III)$ through the interface ($x = b$) between the region II and III can be obtained easily. Therefore, the relationships of these expansion coefficients denoting the incoming and outing waves can be determined by the total scattering matrix S^T

$$\begin{pmatrix} A^{III} \\ B^{I} \end{pmatrix} = S^T \begin{pmatrix} A^{I} \\ B^{III} \end{pmatrix} = \begin{pmatrix} S^T_{11} & S^T_{12} \\ S^T_{21} & S^T_{22} \end{pmatrix} \begin{pmatrix} A^{I} \\ B^{III} \end{pmatrix}. \quad (4.52)$$

In terms of $S(I, II)$, $S(II, II)$, and $S(II, III)$, S^T in the quantum structure shown in Fig. 4.1 can be constructed by

$$S^T = S(I, II) \otimes S(II, II) \otimes S(II, III). \quad (4.53)$$

Thus, here it is imperative to demonstrate how to calculate the elements of the total scattering matrix. Now one considers the two successive scattering matrices defined by

$$S_1 = \begin{pmatrix} r_1 & t_1' \\ t_1 & r_1' \end{pmatrix}, \qquad S_2 = \begin{pmatrix} r_2 & t_2' \\ t_2 & r_2' \end{pmatrix}. \quad (4.54)$$

The resultant scattering matrix of the two matrices is

$$S_{12} = S_1 \otimes S_2 = \begin{pmatrix} r_{12} & t_{12}' \\ t_{12} & r_{12}' \end{pmatrix} \quad (4.55)$$

with

$$\begin{aligned} r_{12} &= r_1 + t_1' r_2 (1 - r_1' r_2) t_1, \\ t_{12}' &= t_1' [r_2 (1 - r_1' r_2)^{-1} r_1' t_2' + t_2'], \\ t_{12} &= t_2 (1 - r_1' r_2)^{-1} t_1, \\ r_{12}' &= t_2 (1 - r_1' r_2)^{-1} r_1' t_2' + t_2'. \end{aligned} \quad (4.56)$$

By using these relations iteratively, expressed in Eq. 4.56, we can easily get the total scattering matrix S^T. In the calculation of the transmission coefficient, we can regard that only the transmitted components are presented in the outgoing lead, i.e., B^{III}=0. According to Eq. 4.52, this leads to

$$A^{III} = S^T_{11} A^{I} \quad (4.57)$$

with A^I being an $N^I \times 1$ vector with elements 1. Then, the transmission coefficient from the mode m in the incoming lead to the mode n in the outgoing lead through all the scattering interfaces is describe as

$$\tau_{mn} = |t_{mn}|^2 \frac{k_n^{III}}{k_m^I} \tag{4.58}$$

where

$$|t_{mn}| = |S_{11}^T|_{mn} \tag{4.59}$$

and k_m^I, k_n^{III} is given by Eq. 4.31. Now summing over all the propagating modes n in the outgoing lead, one gets the transmission coefficient τ_m for the individual incoming mode m at frequency

$$\tau_m(\omega) = \sum_{n,\omega_n<\omega} \tau_{mn} \tag{4.60}$$

with $\omega_n = n\pi/W_{III}$ denoting the cutoff frequency of mode n in the outgoing lead. Thus, the total transmission coefficient is easily obtained by the linear summation of the transmission coefficient of each mode.

$$\tau(\omega)= \sum_{m,\omega_m<\omega} \tau_m(\omega). \tag{4.61}$$

However, for the coupled P-SV waves in the x–y plane, their displacement determined by Eqs. 4.23 and 4.24 can be obtained by solving Eqs. 4.8 and 4.11 under the appropriate boundary conditions. As shown by Li *et al.* [58], the mixed boundary conditions can be expressed by $u_y = T_{xy} = T_{zy} = 0$ at the edges with T_{xy} and T_{zy} denoting the stress tensors, and are considered here. Under these conditions, the displacements in each region ξ (ξ = I, II and III) can be expressed as

$$u_x^\xi = \sum_{n=0}^{N^\xi} \left[\left(A_{Pn}^\xi e^{ik_{Pn}^\xi(x-x_\xi)} - B_{Pn}^\xi e^{-ik_{Pn}^\xi(x-x_\xi)} \right) (k_{Pn}^\xi) \eta_n^\xi(y) \right.$$
$$\left. + \left(A_{SVn}^\xi e^{ik_{SVn}^\xi(x-x_\xi)} + B_{SVn}^\xi e^{-ik_{SVn}^\xi(x-x_\xi)} \right) \left(\frac{n\pi}{W_\xi} \right) \eta_n^\xi(y) \right] \tag{4.62}$$

$$u_y^\xi = \sum_{n=0}^{N^\xi} \left[\left(A_{Pn}^\xi e^{ik_{Pn}^\xi(x-x_\xi)} + B_{Pn}^\xi e^{-ik_{Pn}^\xi(x-x_\xi)} \right) \left(-\frac{n\pi}{W_\xi} \right) \chi_n^\xi(y) \right.$$
$$\left. + \left(A_{SVn}^\xi e^{ik_{SVn}^\xi(x-x_\xi)} - B_{SVn}^\xi e^{-ik_{SVn}^\xi(x-x_\xi)} \right) (k_{SVn}^\xi) \chi_n^\xi(y) \right] \tag{4.63}$$

where

$$\eta_n^{\xi}(y) = \begin{cases} \sqrt{2/W_{\xi}}\cos(n\pi y/W_{\xi}), & \text{if } n \neq 0 \\ \sqrt{1/W_{\xi}}, & \text{if } n = 0 \end{cases} \tag{4.64}$$

$$\chi_n^{\xi}(y) = \begin{cases} \sqrt{2/W_{\xi}}\sin(n\pi y/W_{\xi}), & \text{if } n \neq 0 \\ 0, & \text{if } n = 0 \end{cases} \tag{4.65}$$

$$k_{Pn}^{\xi} = \sqrt{\frac{\omega^2 - (\omega_{Pn}^{\xi})^2}{v_L^2}} \tag{4.66}$$

$$k_{SVn}^{\xi} = \sqrt{\frac{\omega^2 - (\omega_{SVn}^{\xi})^2}{v_T^2}} \tag{4.67}$$

with $\omega_{Pn}^{\xi}=n\pi v_L/W_{\xi}$ and $\omega_{SVn}^{\xi}=n\pi v_T/W_{\xi}$. From Eqs. 4.62 and 4.63, it is seen that there exist four expansion coefficients $(A_P^{\xi}, B_P^{\xi}, A_{SV}^{\xi}, B_{SV}^{\xi})$ in each region corresponding to the transmitted and reflected components of the P and SV waves, respectively. Thus, when the P wave solely goes into the waveguide, the reflection at the interfaces may lead to the mode conversion, namely its reflected wave and the transmitted wave may simultaneously contain both P and SV waves. For the SV wave incidence, the situation is similar. To determine the relationships of these expansion coefficients, we impose the continuity condition of the displacement and stress at all the interfaces ($x = 0, b$) between adjacent regions, i.e.,

$$u_x^{I(II)} = u_x^{II(III)} \tag{4.68}$$

$$u_y^{I(II)} = u_y^{II(III)} \tag{4.69}$$

$$(\lambda + 2\mu)\frac{\partial u_x^{I(II)}}{\partial x} + \lambda\frac{\partial u_y^{I(II)}}{\partial y} = (\lambda + 2\mu)\frac{\partial u_x^{II(III)}}{\partial x} + \lambda\frac{\partial u_y^{II(III)}}{\partial y} \tag{4.70}$$

$$\mu\frac{\partial u_x^{I(II)}}{\partial y} + \mu\frac{\partial u_y^{I(II)}}{\partial x} = \mu\frac{\partial u_x^{II(III)}}{\partial y} + \mu\frac{\partial u_y^{II(III)}}{\partial x}. \tag{4.71}$$

Firstly, we consider the interface ($x = 0$) between regions I and II. By substituting Eqs 4.62 and 4.63 into Eqs. 4.68–4.71, we can easily obtain the following expression:

$$\sum_{n=0}^{N^I}\left[(A^I_{Pn}-B^I_{Pn})k^I_{Pn}\eta^I_n(y)+(A^I_{SVn}+B^I_{SVn})\left(\frac{n\pi}{W_I}\right)\eta^I_n(y)\right]$$

$$=\sum_{n=0}^{N^{II}}\left[(A^{II}_{Pn}-B^{II}_{Pn})k^{II}_{Pn}\eta^{II}_n(y)+(A^{II}_{SVn}+B^{II}_{SVn})\left(\frac{n\pi}{W_{II}}\right)\eta^{II}_n(y)\right],\tag{4.72}$$

$$\sum_{n=0}^{N^I}\left[(A^I_{Pn}+B^I_{Pn})\left(-\frac{n\pi}{W_I}\right)\chi^I_n(y)+(A^I_{SVn}-B^I_{SVn})k^I_{SVn}\chi^I_n(y)\right]$$

$$=\sum_{n=0}^{N^{II}}\left[(A^{II}_{Pn}+B^{II}_{Pn})\left(-\frac{n\pi}{W_{II}}\right)\chi^{II}_n(y)+(A^{II}_{SVn}-B^{II}_{SVn})k^{II}_{SVn}\chi^{II}_n(y)\right],\tag{4.73}$$

$$\begin{aligned}
&\sum_{n=0}^{N^I}(\lambda+2\mu)\left[-(A^I_{Pn}+B^I_{Pn})(k^I_{Pn})^2\eta^I_n(y)\right.\\
&\quad\left.+(B^I_{SVn}-A^I_{SVn})k^I_{SVn}\frac{n\pi}{W_I}\eta^I_n(y)\right]\\
&\quad+\sum_{n=0}^{N^I}\lambda\left[-(A^I_{Pn}+B^I_{Pn})\left(\frac{n\pi}{W_I}\right)^2\eta^I_n(y)\right.\\
&\quad\left.+(A^I_{SVn}-B^I_{SVn})k^I_{SVn}\frac{n\pi}{W_I}\eta^I_n(y)\right]\\
&=\sum_{n=0}^{N^{II}}(\lambda+2\mu)\left[-(A^{II}_{Pn}+B^{II}_{Pn})(k^{II}_{Pn})^2\eta^{II}_n(y)\right.\\
&\quad\left.+(B^{II}_{SVn}-A^{II}_{SVn})k^{II}_{SVn}\frac{n\pi}{W_{II}}\eta^{II}_n(y)\right]\\
&\quad+\sum_{n=0}^{N^{II}}\lambda\left[-(A^{II}_{Pn}+B^{II}_{Pn})\left(\frac{n\pi}{W_{II}}\right)^2\eta^{II}_n(y)\right.\\
&\quad\left.+(A^{II}_{SVn}-B^{II}_{SVn})k^{II}_{SVn}\frac{n\pi}{W_{II}}\eta^{II}_n(y)\right],
\end{aligned}\tag{4.74}$$

$$\sum_{n=0}^{N^I} \mu \left[(A^I_{Pn} - B^I_{Pn}) k^I_{Pn} \left(-\frac{n\pi}{W_I} \right) \chi^I_n(y) + (A^I_{SVn} + B^I_{SVn})(-) \left(\frac{n\pi}{W_I} \right)^2 \chi^I_n(y) \right]$$

$$+ \sum_{n=0}^{N^I} \mu \left[(A^I_{Pn} - B^I_{Pn}) k^I_{Pn} \left(-\frac{n\pi}{W_I} \right) \chi^I_n(y) + (A^I_{SVn} + B^I_{SVn}) (k^I_{SVn})^2 \chi^I_n(y) \right]$$

$$= \sum_{n=0}^{N^{II}} \mu \left[(A^{II}_{Pn} - B^{II}_{Pn}) k^{II}_{Pn} \left(-\frac{n\pi}{W_{II}} \right) \chi^{II}_n(y) + (A^{II}_{SVn} + B^{II}_{SVn})(-) \left(\frac{n\pi}{W_{II}} \right)^2 \chi^{II}_n(y) \right]$$

$$+ \sum_{n=0}^{N^{II}} \mu \left[(A^{II}_{Pn} - B^{II}_{Pn}) k^{II}_{Pn} \left(-\frac{n\pi}{W_{II}} \right) \chi^{II}_n(y) + (A^{II}_{SVn} - B^{II}_{SVn}) (k^{II}_{SVn})^2 \chi^{II}_n(y) \right]. \quad (4.75)$$

Multiplying the transverse wave function $\eta^I_m(y)$ and $\chi^I_m(y)$ in the two sides of Eqs. 4.72 and 4.73, respectively, and $\eta^{II}_m(y)$ and $\chi^{II}_m(y)$ in the two sides of Eqs. 4.74 and 4.75, respectively, after integrating these equations and making use of $\int_0^{W_\xi} \eta^\xi_m(y)\eta^\xi_n(y)dy = \delta_{mn}$ and $\int_0^{W_\xi} \chi^\xi_m(y)\chi^\xi_n(y)dy = \delta_{mn}$, we can obtain

$$(A^I_{Pm} - B^I_{Pm}) k^I_{Pm} + (A^I_{SVm} + B^I_{SVm}) \left(\frac{m\pi}{W_I} \right) - \sum_{n=0}^{N^{II}} D_{mn} \left[(A^{II}_{Pn} - B^{II}_{Pn}) k^{II}_{Pn} + (A^{II}_{SVn} + B^{II}_{SVn}) \left(\frac{n\pi}{W_{II}} \right) \right], \quad (4.76)$$

$$(A^I_{Pm} + B^I_{Pm}) \frac{-m\pi}{W_I} + (A^I_{SVm} - B^I_{SVm}) k^I_{SVm} = \sum_{n=0}^{N^{II}} F_{mn} \left[(A^{II}_{Pn} + B^{II}_{Pn}) \frac{-n\pi}{W_{II}} + (A^{II}_{SVn} - B^{II}_{SVn}) k^{II}_{SVn} \right], \quad (4.77)$$

$$\sum_{m=0}^{N^I}(\lambda+2\mu)D_{nm}\left[-(A^I_{Pm}+B^I_{Pn})(k^I_{Pm})^2+(B^I_{SVm}-A^I_{SVm})k^I_{SVm}\frac{m\pi}{W_I}\right]$$
$$+\sum_{m=0}^{N^I}\lambda D_{nm}\left[-(A^I_{Pm}+B^I_{Pm})\left(\frac{m\pi}{W_I}\right)^2+(A^I_{SVm}-B^I_{SVm})k^I_{SVm}\frac{m\pi}{W_I}\right]$$
$$=(\lambda+2\mu)\left[-(A^{II}_{Pn}+B^{II}_{Pn})(k^{II}_{Pn})^2+(-A^{II}_{SVn}+B^{II}_{SVn})k^{II}_{SVn}\frac{n\pi}{W_{II}}\right]$$
$$+\lambda\left[-(A^{II}_{Pn}+B^{II}_{Pn})\left(\frac{n\pi}{W_{II}}\right)^2+(A^{II}_{SVn}-B^{II}_{SVn})k^{II}_{SVn}\frac{n\pi}{W_{II}}\right], \quad (4.78)$$

$$\sum_{m=0}^{N^I}\mu F_{nm}\left[(A^I_{Pm}-B^I_{Pm})k^I_{Pm}\left(-\frac{m\pi}{W_I}\right)+(A^I_{SVm}+B^I_{SVm})(-)\left(\frac{m\pi}{W_I}\right)^2\right]$$
$$+\sum_{m=0}^{N^I}\mu F_{nm}\left[(A^I_{Pm}-B^I_{Pm})k^I_{Pm}\left(-\frac{m\pi}{W_I}\right)+(A^I_{SVm}+B^I_{SVm})(k^I_{SVm})^2\right]$$
$$=\mu\left[(A^{II}_{Pn}-B^{II}_{Pn})k^{II}_{Pn}\left(-\frac{n\pi}{W_{II}}\right)+(A^{II}_{SVn}+B^{II}_{SVn})(-)\left(\frac{n\pi}{W_{II}}\right)^2\right]$$
$$+\mu\left[(A^{II}_{Pn}-B^{II}_{Pn})k^{II}_{Pn}\left(-\frac{n\pi}{W_{II}}\right)+(A^{II}_{SVn}-B^{II}_{SVn})(k^{II}_{SVn})^2\right] \quad (4.79)$$

with $D_{mn}=\int_0^{W_I}\eta^I_m(y)\eta^{II}_n(y)dy$ and $F_{mn}=\int_0^{W_I}\chi^I_m(y)\chi^{II}_n(y)dy$. Note that here D_{mn} is the same as that presented in Eqs. 4.39–4.43 for the SH wave. Similarly, we can also get the explicit form of F_{mn} in terms of Eq. 4.65

$$F_{mn}=0, \quad \text{if } m=0 \text{ or } n=0; \quad (4.80)$$

$$F_{mn}=\sqrt{\frac{W_I}{W_{II}}}, \quad \text{if } m\neq 0,\ n\neq 0,\ \frac{m}{W_I}=\frac{n}{W_{II}}; \quad (4.81)$$

$$F_{mn}=\frac{1}{\sqrt{W_I W_{II}}}\left[\frac{-1}{\frac{m\pi}{W_I}+\frac{n\pi}{W_{II}}}\sin\left(m\pi+\frac{n\pi W_I}{W_{II}}\right)\right.$$
$$\left.+\frac{1}{\frac{m\pi}{W_I}+\frac{n\pi}{W_{II}}}\sin\left(m\pi-\frac{n\pi W_I}{W_{II}}\right)\right], \quad \text{if } m\neq 0,\ n\neq 0,\ \frac{m}{W_I}\neq\frac{n}{W_{II}}. \quad (4.82)$$

By using the similar scheme, we can also deal with the case at the interface ($x = b$) between regions II and III. Then, similar to the case for the SH wave elaborated above, we rewrite all the equations into the form of the matrix. Thus, the total scattering matrix for the coupled P-SV wave in a given structure shown in Fig. 4.1 can be defined by

$$\begin{pmatrix} A_P^{III} \\ A_{SV}^{III} \\ B_P^{I} \\ B_{SV}^{I} \end{pmatrix} = S^T \begin{pmatrix} A_P^{I} \\ A_{SV}^{I} \\ B_P^{III} \\ B_{SV}^{III} \end{pmatrix} = \begin{pmatrix} S_{11}^T & S_{12}^T & S_{13}^T & S_{14}^T \\ S_{21}^T & S_{22}^T & S_{23}^T & S_{24}^T \\ S_{31}^T & S_{32}^T & S_{33}^T & S_{34}^T \\ S_{41}^T & S_{42}^T & S_{43}^T & S_{44}^T \end{pmatrix} \begin{pmatrix} A_P^{I} \\ A_{SV}^{I} \\ B_P^{III} \\ B_{SV}^{III} \end{pmatrix}, \qquad (4.83)$$

where A_P^I, B_P^I, A_{SV}^I, B_{SV}^I, A_P^{III}, B_P^{III}, A_{SV}^{III}, and B_{SV}^{III} are the column vectors with the corresponding elements A_{Pm}^I, B_{Pm}^I, A_{SVm}^I, B_{SVm}^I, A_{Pn}^{III}, B_{Pn}^{III}, A_{SVn}^{III}, and B_{SVn}^{III}. Because these two waves generally couple with each other in the interface scattering process, the transmission coefficient should be discriminated according to the P, SV incoming and P, SV outgoing waves, respectively. For the sake of unification, here we denote them as $\tau_{\alpha m,\beta n}$, which represents the transmission coefficient from the m mode of $\alpha(\alpha=P,\ SV)$ wave to the n mode of $\beta(\beta=P,\ SV)$ wave. Thus,

$$\tau_{Pm,Pn}=(|S_{11}^T|_{mn})^2\frac{k_{Pn}^{III}}{k_{Pm}^{I}}, \qquad \tau_{Pm,SVn}=(|S_{21}^T|_{mn})^2\frac{k_{SVn}^{III}}{k_{Pm}^{I}}, \qquad (4.84)$$

if the incident wave is P wave, and

$$\tau_{SVm,Pn}=(|S_{12}^T|_{mn})^2\frac{k_{Pn}^{III}}{k_{SVm}^{I}}, \qquad \tau_{SVm,SVn}=(|S_{22}^T|_{mn})^2\frac{k_{SVn}^{III}}{k_{SVm}^{I}}, \qquad (4.85)$$

if the incident wave is SV wave. In these equations above, k_{Pm}^{ξ}, k_{Pn}^{ξ}, k_{SVm}^{ξ}, and k_{SVn}^{ξ} are the corresponding wave vector determined by Eqs. 4.66 and 4.67, respectively. Thus, summing over the transmitted modes coupled to the incoming modes, the transmission coefficient of the individual incoming m mode for a given ω is expressed as

$$\tau_{Pm} = \sum_{n,\omega_{Pn}<\omega} \tau_{Pm,Pn} + \sum_{n,\omega_{SVn}<\omega} \tau_{Pm,SVn} \qquad (4.86)$$

for the P wave and

$$\tau_{SVm} = \sum_{n,\omega_{Pn}<\omega} \tau_{SVm,Pn} + \sum_{n,\omega_{SVn}<\omega} \tau_{SVm,SVn} \qquad (4.87)$$

for the SV wave. In these expressions, $\omega_{Pn} = n\pi v_L/W_{III}$ and $\omega_{SVn} = n\pi v_T/W_{III}$ denote the cutoff frequency of the P and SV wave in the outgoing lead, respectively. Thus, the total transmission coefficient can be expressed as

$$\tau = \sum_{m,\omega_{Pm}<\omega} \tau_{Pm} + \sum_{m,\omega_{SVm}<\omega} \tau_{SVm} \tag{4.88}$$

with $\omega_{Pm} = m\pi v_L/W_{III}$ and $\omega_{SVm} = m\pi v_T/W_{III}$.

We stress that in this formalism, once we know how to handle the simple T-shaped structure, the complicated problem can be also done similarly. For instance, for the complicated quantum systems such as the rough surface, catenoidal contacts, and so on, we firstly subdivide all the system into very small segments so that each segment can be regarded to have a uniform cross area. Then one tries to calculate the scattering matrices of arbitrary two successive segments by using the above procedures, so as to obtain the total scattering matrix connecting the incoming lead to the outgoing lead. As long as the length of each segment is small enough, such the scheme is very consistent with analytic calculations, which is also confirmed by Tanaka *et al.* [65]. In the following section, we will numerically describe the properties of low temperature ballistic thermal transport by phonons in low-dimensional quantum structures.

4.3 Properties of Low-Temperature Ballistic Thermal Transport by Phonons in Low-Dimensional Quantum Structures

4.3.1 *Properties of Ballistic Thermal Transport in 2D Quantum Structures*

Here, we consider a 2D quantum system in the x–y plane, in which the inhomogeneities are embedded between two perfect semi-infinite 2D quantum wires. In such a structure, we usually divide the entire structure into three regions: the central scattering region and the two lead regions, which are connected separately to the hot and cold reservoirs. It is assumed that the contacts between the two leads and reservoirs are perfect and the elastic scattering for

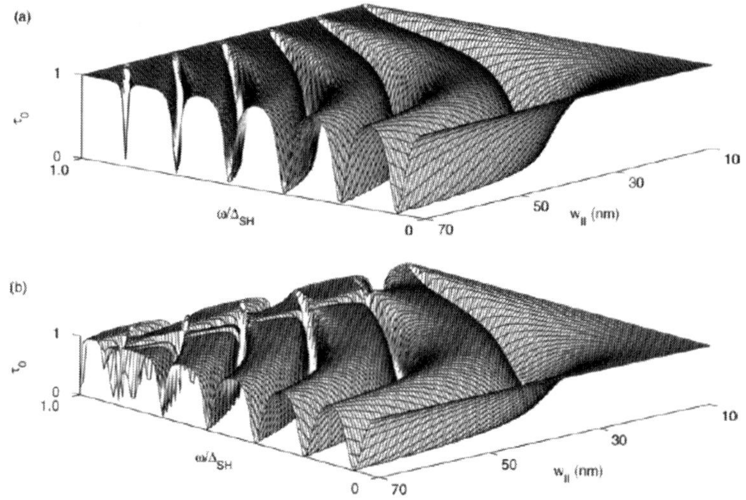

Figure 4.2 Transmission coefficient *vs.* the transverse width W_{II} and the frequency ω/Δ_{SH} for the T-shaped quantum structure with $W_I = 10$ nm. Here, $\omega_{m+1} - \omega_m = \pi v_T / W_I$ represents the splitting of the cutoff frequency between the $(m + 1)$ mode and the m mode [58]. (a) $b = 10$ nm; (b) $b = 12$ nm.

phonons stems from inhomogeneities in central region. Here, two typical of quantum structures with inhomogeneities are considered: (i) T-shaped quantum waveguide; (ii) single structural defect such as the defect being consisted of the void and clamped material located in the midsection of nanowires. Some general properties are revealed.

For case (i), it was demonstrated by Li *et al.* [58] that how several lowest SH modes go through the T-shaped quantum structure. Fig. 4.2 firstly shows the transmission coefficient of the lowest SH mode (zero mode) as a function of the incident frequency and the transverse width W_{II}. It is seen that the zero mode with the cutoff frequency $\omega = 0$ can propagate through such the structure, which is substantially different from the case for electron transport. This is due to the fact that the acoustic phonon satisfies the stress-free boundary condition, while the electron satisfies the hard wall boundary condition. When $\omega \rightarrow 0$, all the transmission coefficient approaches unity, which is consistent with that for the defective graphene ribbons calculated by the non-equilibrium Green's function [52]. This is because that the long wavelength phonons with $\omega \to 0$ are hardly scattered by inhomogeneities in T-shaped quantum structure. This will lead to an important conclusion that all the thermal conductance has the universal quantum values in the limit $T \to 0$, regardless of the geometric parameters of inhomogeneities. When the frequency is increased, the transmission coefficient is reduced due to the interface scattering, and such the reduction is more evident for larger W_{II}. From Fig. 4.2a, it is also seen that for a given frequency, the phonon transmission exhibits a periodic pattern as a function of the transverse width of T-shaped structure. However, for larger b (Fig. 4.2b), when the frequency is added up to a certain degree where the $1th$ mode can be just excited in the T-shaped structure, the periodic pattern of the transmission coefficient is destroyed strongly with some irregular oscillations. This is mainly due to the coupling effect between the higher modes being excited in the scattering region and the incident zero mode. In general, the more modes couple with the incident mode, the more complex is the transmission coefficient. As a result, the transmission coefficient becomes more complicated with increasing the frequency (not shown). An important feature in Fig. 4.2 is that the transmission coefficient presents many dips at certain frequency, which are clearly distinguished from the dips originating from the van singularity of the phonon branches. Especially, some dips are of $\tau_0 = 0$ indicating that the resonant reflection behavior occurs in this case. Similar phenomenon can also be observed for the electron transport [86] or for higher order phonon modes $(m \geq 1)$ [57]. As discussed above, the resonant reflection can be understood as a result of the resonant coupling between the incoming zero mode and many higher order modes being excited in T-shaped structure.

In terms of Eq. 4.3, one can study and discuss the thermal conductance under the influence of external factors, i.e., temperature and structural parameters, as also shown by Li *et al.* [58]. For the T-shaped junction example, although the transmission coefficient τ_m has many irregular oscillations as mentioned above, these oscillations can be not observed in the thermal conductance, and instead the overall thermal conductance increase monotonously with increasing the temperature. This is due to the fact that the wide frequency integration range in Eq. 4.3 can smear out all the oscillations in the transmission coefficient, which is confirmed here. Clearly in the low temperature region, the thermal conductance is contributed by the several lowest phonon modes, especially the zero acoustic mode. As discussed above, all the thermal conductance approaches the universal quantum value in the limit $T \to 0$ due to the perfect transmission of zero mode at $\omega \to 0$. Note that the reduced thermal conductance K/T of the zero mode seems to be insensitive to the variation of temperature, while that for higher modes rapidly increases as the temperature goes up. In addition, with the increase of the width or length of T-shaped junction, the overall reduced thermal conductance K/T decreases noticeably, which indicates that the thermal conductance may be effectively tuned by adjusting the geometric parameters of T-shaped junction.

For case (ii), Chen *et al.* [51] found that the dependence of the thermal conductance on the temperature is qualitatively different for different types of defects. When the defect is void (see Fig. 4.3), namely the stress-boundary condition is applied between the defect and surroundings, it is shown that only the zero mode contributes to the thermal conductance and K approaches the universal quantum value $\pi^2 k_B^2 T/3h$ as the temperature $T \to 0$. Zero mode with $m = 0$ propagating through the structure is the peculiarity of the elastic wave, which stems from the stress-free boundary condition employed for the void defect allowing the propagation of the mode with $\omega = 0$. Moreover, the reduced thermal conductance K/T in the limit $T \to 0$ is independent of the size of the defect (Fig. 4.3(a)), which indicates the scattering effect of the void defect for the long wavelength phonons with $\omega \to 0$ is very small. By increasing the temperature, because many higher modes being excited begin to contribute to the thermal conductance, their reduced thermal conductance K/T is monotonously increased indicating $K \propto T^x (x > 1)$ for higher modes (Fig. 4.3b and c). However, for the zero mode the reduced thermal conductance K/T unexpectedly decreases with increasing temperature indicating $K \propto T^x (0 < x < 1)$ for the zero mode. Due to the main contribution of the zero mode to the thermal conductance, the decrease of the overall reduced thermal conductance K/T can be also found in low

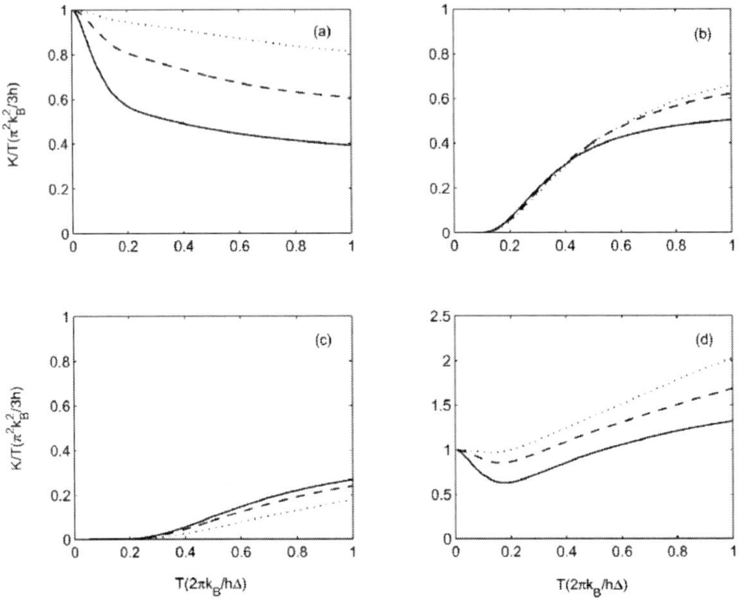

Figure 4.3 The thermal conductance divided by temperature K/T, which is reduced by the zero-temperature universal value $\pi^2 k_B^2/3h$, as a function of the reduced temperature $k_B T/\hbar\Delta$ for different structural parameters, where $\Delta = \omega_{m+1} - \omega_m = \pi v_T/W_I$. Here, the defect is void. (a)–(c) correspond to K/T of modes 0, 1, and 2, respectively, and (d) corresponds to the total K/T. Please refer to Ref. [51] for further details.

temperature (Fig. 4.3d), which seems consistent with experimental results. Moreover, the reduced thermal conductance K/T of the zero mode decreases with the increase of the thickness of the void defect; while the case is just the reverse for higher modes at a certain temperature. These indicate the effects of the void defect on the thermal conductance sensitively depend on the index of the modes.

When the defect consists of clamped material, namely the hard wall boundary condition should be satisfied between the defect and surroundings, it is seen that these results shown in Fig. 4.4 are obviously distinguishing from that for the void defect. Firstly, in the limit $T \rightarrow 0$, the quantum thermal conductance is not observed. There exists a threshold temperature where the zero

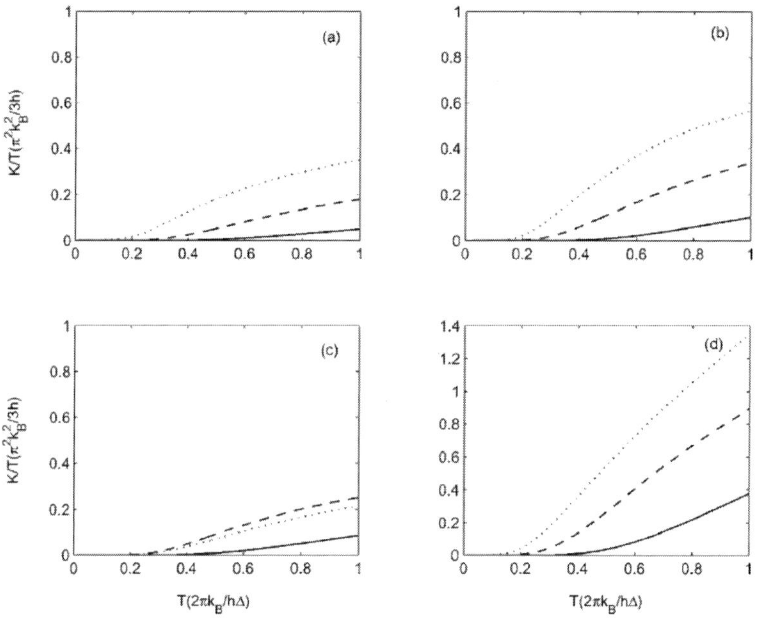

Figure 4.4 The thermal conductance divided by temperature K/T, which is reduced by the zero-temperature universal value $\pi^2 k_B^2/3h$, as a function of the reduced temperature $k_B T/\hbar\Delta$ for different structural parameters, where $\Delta = \omega_{m+1} - \omega_m = \pi v_T/W_I$. Here, the defect is clamped material [51].

mode can be excited, and the reduced thermal conductance for the zero mode monotonously increase with increasing temperature, similar to the behaviors for higher modes, such as modes 1 and 2. This result results from the hard wall boundary condition between the defect and surroundings. Moreover, due to the direct increase of the reduced thermal conductance for the zero mode, the decrease of the overall reduced thermal conductance K/T can not be observed in low temperature region (Fig. 4.4d). Thus, one knows that the universal quantum thermal conductance can be observed only under condition of the ballistic boundary, where the zero mode can be excited in the limit $T \rightarrow 0$. Moreover, the effects of defects on the thermal conductance depend upon the type and sizes of the defects.

4.3.2 *Ballistic Thermal-Transport Properties in 2D Three-Terminal Quantum Structures*

Note that two preceding cases only consider the ballistic phonon transport in the two-terminal dielectric systems. Recently, various three-terminal junctions have emerged as building blocks in the formation of nanoscale electronic devices. Some applications of three-terminal junctions are proposed based on the theoretically investigated properties of electron transport [87, 88]. For the electronic transport in three-terminal system, it has become well known that the third terminal can be used to apply a gate voltage and thus compatibly control the electric current in the channel built by the two other terminals by using the gate as a voltage probe [87]. More recently, phonon transport properties in three-terminal systems have drawn increasing attentions due to the potential applications in thermal devices. As shown in Fig. 4.5, for an asymmetric y-branch three-terminal junction [89], one finds that in the low frequency region, where the two lowest phonon modes can be excited, there exist obvious mode splitting behaviors in such structure. For instance, the incoming zero mode mostly transports into an outgoing lead with the same thickness to the incident lead (namely region B), while the 1*th* mode mostly enters into the other lead (namely region A). Similar mode splitting behaviors can be also observed in this case for the higher incident frequency. These interesting properties can be well understood as a result of the resonant coupling between the incident lead and different outgoing leads. In addition, Ming *et al.* presented a study of the thermal properties of three-terminal mesoscopic dielectric systems in the nonlinear response regime at low temperature [90]. It is demonstrated that when temperatures T_{L} and T_{R} are applied to the left and right lead of a symmetric three-terminal system in push-pull fashion with $T_{\mathrm{L}} \neq T_{\mathrm{R}}$, the temperature of the central lead is always higher than $(T_{\mathrm{L}} + T_{\mathrm{R}})/2$. Similar phenomenon can be also observed for an asymmetric three-terminal system. According to Eq. 4.2, these are analytically explained by making the Taylor expansion for the Bose–Einstein distribution function. There should be no thermal rectification in two-terminal mesoscopic dielectric system, since the transmission coefficient of ballistic

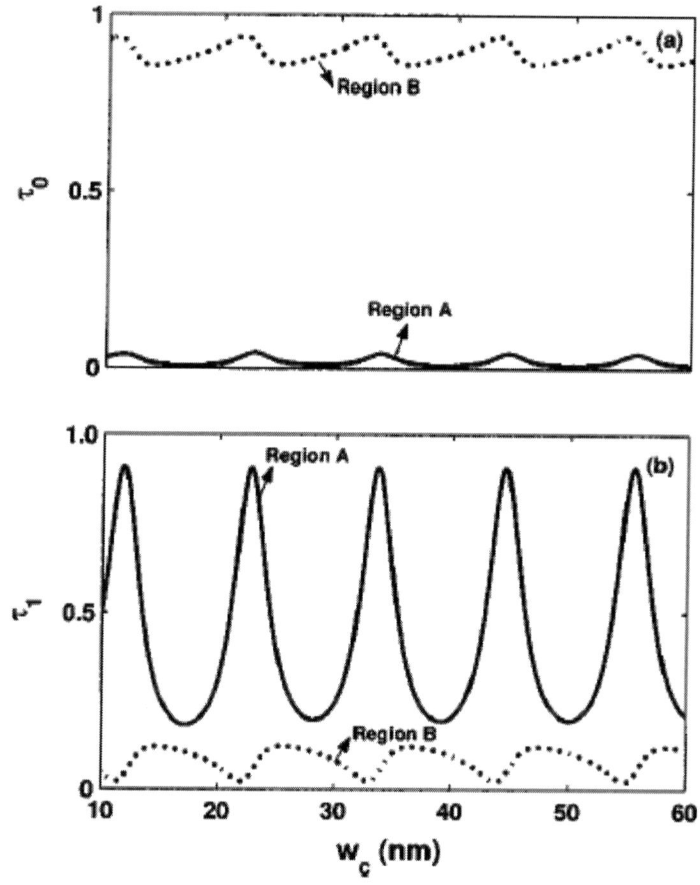

Figure 4.5 Transmission coefficient of (a) mode 0 and (b) mode 1 vs W_C for asymmetric y-branch three-terminal junction. The incident frequency is $1.1\pi v_T / W_I$. Solid and dotted curves correspond to the transmission in regions A and B, respectively [89].

phonons is independent of the temperature. However, the three-terminal systems may offer a proper option for ballistic thermal rectification by introducing incoherence to the phonon transmission from the third terminal, which is confirmed by Zhang *et al.* [91]. They found that ballistic thermal rectification exists in asymmetric three-terminal atomic nanojunctions, even in symmetric three-terminal structures with spin-phonon interaction. More recently, the ballistic thermal rectification in a similar three-terminal mesoscopic dielectric system at low temperature is analytically obtained based on the Landauer transport theorem, and the rectifying efficiency can be efficiently tuned by changing the structural asymmetry or adjusting the central probe [92]. These results indicate that three-terminal mesoscopic junctions have an important application in nanodevices, such as mode splitter, thermal rectifier, and so on.

4.3.3 *Properties of Ballistic Thermal Transport in 3D Quantum Structures*

It is worthy noting that these previous studies mainly focused on the ballistic thermal transport in the quantum structures with inhomogeneities, where it is assumed that the thickness of each region is very small with respect to the other dimensions and also to the elastic phonon wavelength. In fact, the out-of-plane dynamics is ignored and only the 2D geometries are considered. By developing the 2D model technique, Li *et al.* [47] first calculated the ballistic thermal conductance in a 3D quantum structure with an abrupt junction, and predicted that the thermal conductance is lower than the universal unit $\pi^2 k_B^2 T/3h$ and is strongly influenced by the abrupt junction structure in the low temperature limit. Recently, Peng *et al.* presented a numerical study of the low temperature thermal conductance in a 3D quantum wire with a stub structure under two different boundary conditions [56]. The results in Fig. 4.6 show that when stress-free boundary conditions are applied, the thermal conductance *vs.* temperature displays similar behavior as that in 2D model. In the low temperature limit, the thermal conductance approaches the quantum value $\pi^2 k_B^2 T/3h$ regardless of the geometric structure, which agrees with the results by 2D model, and also agrees with the experimental results. However, when hard wall boundary conditions are applied, the thermal conductance *vs.* temperature displays different behavior from that for 2D models. More recently, Xie *et al.* investigated the ballistic phonon transport and thermal conductance in a 3D double-bend quantum structure [60]. These studies reveal that the thermal conductance in 3D model is much larger than that in 2D model for the same structural

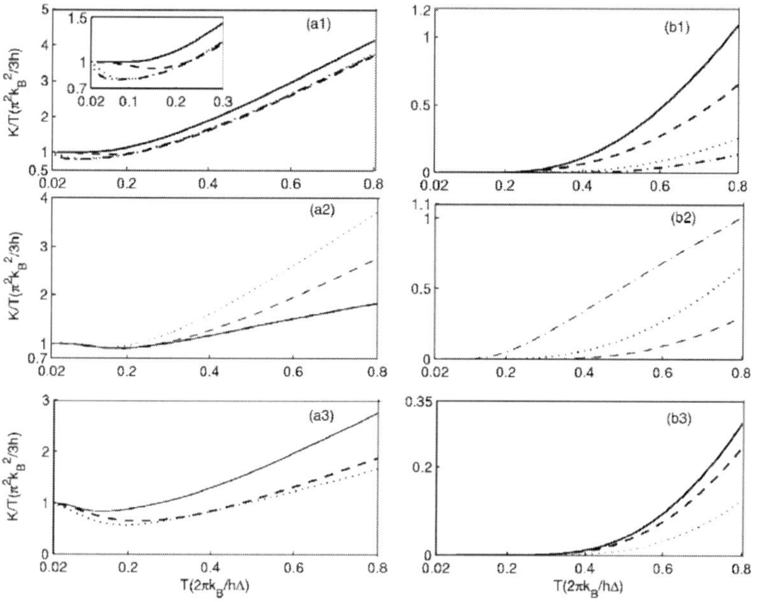

Figure 4.6 Thermal conductance divided by temperature K/T, which is reduced by the zero-temperature universal value $\pi^2 k_B^2/3h$, as a function of the reduced temperature $k_B T/\hbar\Delta$ in a three-dimensional quantum wire with stub structure. (a1–a3) and (b1–b3) correspond to the stress-free and hard wall boundary conditions, respectively. Note that the dash-dotted curve was calculated by the 2D model in (a2) and (b2) [56].

parameters and temperature. Moreover, the dependence of the thermal conductance on the temperature in 3D model is different from that in 2D model. However, some interesting physical effects that appeared in 2D model, such as the universal quantum thermal conductance, resonant transmission, stop-frequency gap, and so on, can be also observed in 3D model. So these results indicate that although there exist some differences of thermal properties between the 2D and 3D models due to the two transverse boundary confinement in 3D model but only one in 2D model. The 2D model can still hold good approximation to describe qualitatively the phonon transmission and thermal conductance in real 3D geometry.

4.3.4 *Ballistic Thermal Transport Contributed by the Coupled P-SV Waves in Low-Dimensional Quantum Structures*

Li *et al.* [58] systematically investigated the ballistic thermal-transport properties of the coupled P-SV waves through a T-shaped quantum structure and provided a detailed comparison between the SH waves and the coupled P-SV waves. It is demonstrated that the coupled P-SV waves show some new transmission characters, different from the case for the SH waves due to the mode conversion and mode mixing effect. For example, when the P wave is incident, zero transmission only occurs at low frequency where only the zero mode of P wave can be excited in such a waveguide. For higher frequency, when the first mode of SV wave starts to be excited, there is a rapid decrease in transmission coefficient. These are inconsistent with the results of the SH wave for the same structural parameters. In addition, compared to the SH wave, in low-frequency region, the transmission coefficient of the P wave becomes more complicated and exhibits more resonance transmission peaks.

4.4 Summary

In the present review, we introduce the elastic continuum model to describe the elastic phonon modes in low-dimensional nanostructures, and present the necessary details of the scattering-matrix method to calculate the transmission coefficient and the Landauer formula for the thermal conductance. Then the effects of inhomogeneities, boundary conditions, dimensions, etc., on the ballistic phonon transmission and thermal conductance are summarized and reviewed. Results show some interesting physical properties, such as the universal quantum thermal conductance, periodic transmission behaviors, stop-frequency gap, and so on. In fact, thermal management has become an essential issue for continuing progress in the electronic industry. The present results provide an efficient way to control the thermal conductance artificially by adjusting the parameters of the proposed nanostructures. These will be helpful to not only comprehensively understand the behaviors of low temperature ballistic thermal transport in low-dimensional structures, but also design some functional thermal devices.

References

1. Wees, B. J. V., Houten, H. V., Beenakker, C. W. J., and Williamson, J. G. (1988) Quantized conductance of point contacts in a two-dimensional electron gas, *Phys. Rev. Lett.*, **60**, pp. 848–850.
2. Szafer, A. and Stone, D. A. (1989) Theory of Quantum conduction through a constriction *Phys. Rev. Lett.* **62**, pp. 300–303.
3. Xu, H. Q. (1995) Electron transport through one-dimensional lateral surface superlattices in magnetic fields, *Phys. Rev. B*, **52**, pp. 5803–5812.
4. Balandin, A. A. (2009) Better computing through CPU cooling, *IEEE Spectrum*, pp. 29–33.
5. Balandin, A. A. and Wang, K. L. (1998) Significant decrease of the lattice thermal conductivity due to phonon confinement in a free-standing semiconductor quantum well, *Phys. Rev. B*, **58**, pp. 1544–1549.
6. Chuang, D. Y., Hogan, T., Brazis, P., Lane, M. R., Kannewurf, C., Bastea, M., Uher, C., and Kanatzidis, M. G. (2000) $CsBi_4Te_6$: a high-performance thermoelectric material for low-temperature applications, *Science*, **287**, pp. 1024–1027.
7. Majumdar, A. (2004) Thermoelectricity in semiconductor nanostructures, *Science*, **303**, pp. 777–778.
8. Chang, C. W., Okawa, D., Majumdar, A., *et al.* (2006) Solid-state thermal rectifier, *Science*, **314**, pp. 1121–1124.
9. Yang, N., Zhang, G. and Li, B. (2008) Carbon nanocone: a promising thermal rectifier, *Appl. Phys. Lett.*, **93**, pp. 243111–243113.
10. Li, B. W., Wang, L., and Casati, G. (2006) Negative differential thermal resistance and thermal transistor, *Appl. Phys. Lett.*, **88**, pp. 143501–143503.
11. Li, B. W., Wang, L., and Casati, G. (2004) Thermal diode: Rectification of heat flux, *Phys. Rev. Lett.*, **93**, pp. 184301–184304.
12. Wang, L. and Li, B. W. (2008) Thermal memory: a storagy of phononic information, *Phys. Rev. Lett.*, **101**, pp. 267203–067206.
13. Tighe, T. S., Worlock, J. W. and Roukes, M. L. (1997) Direct thermal conductance measurements on suspended monocrystalline nanostructures, *Appl. Phys. Lett.*, **70**, pp. 2687–2689.

14. Schwab, K., Henriksen, E. A., Worlock, J. M., *et al.* (2000) Measurement of the quantum of thermal conductance, *Nature*, **404**, pp. 974–977.
15. Chiu, H. Y., Deshpande, V. V., Postma, H. W. C., Lau, C. N., Miko, C. and Bockrath, M. (2005) Ballistic phonon thermal transport in multiwalled carbon nanotubes, *Phys. Rev. Lett.*, **95**, pp. 226101–226104.
16. Balandin, A. A., Ghosh, S., Bao, W. Z., Calizo, I., Desalegne, T., Miao, F., and Lau, C. N. (2008) Superior thermal conductivity of single-layer graphene, *Nano Lett.*, **8**, pp. 902–907.
17. Chang, C. W., Okawa, D., Garcia, H., Majumdar, A. and Zettl, A. (2008) Breakdown of Fourier's law in nanotube thermal conductors, *Phys. Rev. Lett.*, **101**, pp. 075903–075906.
18. Chang, C. W., Fennimore, A. M., Afanasiev, A., Okawa, D., Ikuno, T., *et al.* (2006) Isotope effect on the thermal conductivity of boron nitride nanotubes, *Phys. Rev. Lett.*, **97**, pp. 085901–085904.
19. Chen, R., Hochbaum, A. I., Murphy, P., Moore, J., Yang, P. and Majumdar, A. (2008) Thermal conductance of thin silicon nanowires, *Phys. Rev. Lett.*, **101**, pp. 105501–105504.
20. Rego, L. G. C. and Kirczenow, G. (1998) Quantized thermal conductance of dfielectric quantum wires, *Phys. Rev. Lett.*, **81**, pp. 232–235.
21. Angelescu, D. E., Cross, M. C., and Roukes, M. L. (1998) Heat transport in mesoscopic systems, *Superlattices and Microstructures*, **23**, pp. 673–689.
22. Blencowe, M. P. (1999) Quantum energy flow in mesoscopic dielectric structures, *Phys. Rev. B*, **59**, pp. 4992–4998.
23. Blencowe, M. P. (2004) Quantum electromechanical systems, *Phys. Rev.*, **395**, pp. 159–222.
24. Greiner, A., Reggiani, L., Kuhn, T., and Varani, L. (1997) Thermal conductivity and Lorenz number for one-dimensional ballistic transport, *Phys. Rev. Lett.*, **78**, pp. 1114–1117.
25. Cross, M. C. (2001) Elastic wave transmission at an abrupt junction in a thin plate application to heat transport and vibrations in mesoscopic systems, *Phys. Rev. B*, **64**, pp. 085324–085345.
26. Dhar, A. and Roy, D. (2006) Heat transport in Harmonic lattices, *J. Stat. Phys.*, **125**, pp. 801–820.
27. Chang, C. M. and Geller, M. R. (2005) Mesoscopic phonon transmission through a nanowire-bulk contact, *Phys. Rev. B*, **71**, pp. 125304–125311.
28. Landauer, R. (1970) Electrical resistance of disordered one-dimensional lattices, *Philosophical Magazine*, **21**, pp. 863–867.

29. Buttiker, M. (1986) Four-terminal phase-coherent conductance, *Phys. Rev. Lett.*, **57**, pp. 1761–1764.

30. Markussen, T., Jauho, A. P., and Brandbyge, M. (2008) Heat conductance is strongly anisotropic for pristine silicon nanowires, *Nano Lett.*, **8**, pp. 3771–3775.

31. Markussen, T., Jauho, A. P., and Brandbyge, M. (2009) Electron and phonon transport in silicon nanowires: atomistic approach to thermoelectric properties, *Phys. Rev. B*, **79**, pp. 035415–035421.

32. Hochbaum, A. I., Chen, R., Delgabo, R. D., Liang, W. J., Garnett, E. C., Najarian, M., Majumdar, A. and Yang, P. D. (2008) Enhanced thermoelectric performance of rough silicon nanowires, *Nature*, **451**, pp. 163–168.

33. Yamamoto, T. and Watanabe, K. (2004) Empirical-potential study of phonon transport in graphitic ribbons, *Phys. Rev. B*, **70**, pp. 245402–245408.

34. Hu, J. N., Ruan, X., and Chen, Y. P. (2009) Thermal conductivity and thermal rectification in graphene nanoribbons: a molecular dynamics study, *Nano Lett.*, **9**, pp. 2730–2735.

35. Xu, Y., Chen, X. B., Gu, B. L., and Duan, W. H. (2009) Intrinsic anisotropy of thermal conductance in graphene nanoribbons, *Appl. Phys. Lett.*, **95**, pp. 233116–233118.

36. Xu, Y., Chen, X. B., Wang, J. S., Gu, B. L., and Duan, W. H. (2010) Thermal transport in graphene junctions and quantum dots, *Phys. Rev. B*, **81**, pp. 195425–195431.

37. Lan, J. H., Wang, J. S., Gan, C. K., and Chin, S. K. (2009) Edge effects on quantum thermal transport in graphene nanoribbons: tight-binding calculations, *Phys. Rev. B*, **79**, pp. 115401–115405.

38. Savin, A. V., Kivshar, Y. S. and Hu, B. (2010) Suppression of thermal conductivity in graphene nanoribbons with rough edges, *Phys. Rev. B*, **82**, pp. 195422–195430.

39. Yang, B. and Chen, G. (2003) Partially coherent phonon heat conduction in superlattices, *Phys. Rev. B*, **67**, pp. 195311–195314.

40. Hepplestone, S. P. and Srivastava, G. P. (2010) Theory of interface scattering of phonons in superlattices, *Phys. Rev. B*, **82**, pp. 144303–144308.

41. Zhou, L. P., Wang, M. P., Zhu, J. J., Peng, X. F., and Chen, K. Q. (2009) Effects of dimensionality on the ballistic phonon transport and thermal conductance in nanoscale structures, *J. Appl. Phys.*, **105**, pp. 144318–144324.

42. Yamamoto, T., Watanabe, S., and Watanabe, K. (2004) Universal features of quantized thermal conductance of carbon nanotubes, *Phys. Rev. Lett.*, **92**, pp. 075502–075505.
43. Yamamoto, T. and Watanabe, K. (2006) Nonequilibrium Green's function approach to phonon transport in defective carbon nanotubes, *Phys. Rev. Lett.*, **96**, pp. 255503–255506.
44. Yamamoto, T., Sasaoka, K. and Watanabe, S. (2011) Universality and diversity in a phonon-transmission histogram of isotope-disordered carbon nanotubes, *Phys. Rev. Lett.*, **106**, pp. 215503–215506.
45. Mingo, N. and Broido, D. A. (2005) Carbon nanotube ballistic thermal conductance and its limit, *Phys. Rev. Lett.*, **95**, pp. 096105–096108.
46. Stewart, D. A., Savic, I. and Mingo, N. (2009) First-principles calculations of the isotope effect on boron nitride nanotube thermal conductivity, *Nano Lett.*, **9**, pp. 81–84.
47. Li, W. X., Liu, T. Y. and Liu, C. L. (2006) Phonon transport through a three-dimensional abrupt junction, *Appl. Phys. Lett.*, **89**, pp. 163104–163106.
48. Santamore, D. H. and Cross, M. C. (2001) Effect of phonon scattering by surface roughness on the universal thermal conductance, *Phys. Rev. Lett.*, **87**, pp. 115502–115505.
49. Santamore, D. H. and Cross, M. C. (2001) Effect of surface roughness on the universal thermal conductance, *Phys. Rev. B*, **63**, pp. 184306–184311.
50. Akguc, G. B. and Gong, J. B. (2009) Wave-scattering formalism for thermal conductance in thin wires with surface disorder, *Phys. Rev. B*, **80**, pp. 195408–195414.
51. Chen, K. Q., Li, W. X., Duan, W. H., Shuai, Z. and Gu, B. L. (2005) Effect of defects on the thermal conductivity in a nanowire, *Phys. Rev. B*, **72**, pp. 045422–045426.
52. Xie, Z. X., Chen, K. Q., and Duan, W. H. (2011) Thermal transport by phonons in zigzag graphene nanoribbons with structural defects *J. Phys.: Condens. Matter*, **23**, pp. 315302–315307.
53. Jiang, J. W., Wang, B. S., and Wang, J. S. (2011) First principle study of the thermal conductance in graphene nanoribbon with vacancy and substitutional silicon defects, *Appl. Phys. Lett.*, **98**, pp. 113114–113116.
54. Peng, X. F., Chen, K. Q., Wan, Q., Zou, B. S., and Duan, W. H. (2010) Quantized thermal conductance at low temperature in quantum wire with catenoidal contacts, *Phys. Rev. B*, **81**, pp. 195317–195322.

55. Tang, L. M., Wang, L. L., Chen, K. Q., Huang, W. Q., and Zou, B. S. (2006) Coupling effect on phonon thermal conductance in a double-stub quantum wire, *Appl. Phys. Lett.*, **88**, pp. 163505–163507.
56. Peng, X. F., Chen, K. Q., Zou, B. S., and Zhang, Y. (2007). Ballistic thermal conductance in a three-dimensional quantum wire modulated with stub structure, *Appl. Phys. Lett.*, **90**, pp. 193502–193504.
57. Yang, P., Sun, Q. F., Guo, H., and Hu, B. (2007) Thermal transport in a dielectric T-shaped quantum wire, *Phys. Rev. B*, **75**, pp. 235319–235324.
58. Li, W. X., Chen, K. Q., Duan, W. H., Wu, J., and Gu, B. L. (2004) Acoustic phonon transport through a T-shaped quantum waveguide, *J. Phys.: Condens. Matter*, **16**, pp. 5049–5059.
59. Huang, W. Q., Chen, K. Q., Shuai, Z., Wang, L. L., Hu, W. Y., and Zou, B. S. (2005) Acoustic-phonon transmission and thermal conductance in a double-bend quantum waveguide, *J. Appl. Phys.*, **98**, pp. 093524–093530.
60. Xie, F., Chen, K. Q., Wang, Y. G., Wan, Q., Zou, B. S., and Zhang, Y. (2008) Acoustic phonon transport and ballistic thermal conductance through a three-dimensional double-bend quantum structure, *J. Appl. Phys.*, **104**, pp. 054312–054316.
61. Xie, F., Chen, K. Q., Wang, Y. G., and Zhang, Y. (2008) Effect of the evanescent modes on ballistic thermal transport in quantum structures, *J. Appl. Phys.*, **103**, pp. 084501–084505.
62. Wang, J. S., Wang, J., and Lu, J. T. (2008) Quantum thermal transport in nanostructures, *Eur. Phys. J. B.*, **62**, pp. 381–404.
63. Dubi, Y. and Ventra, M. D. (2011) Colloquium: Heat flow and thermoelectricity in atomic and molecular junctions, *Rev. Mod. Phys.*, **83**, pp. 131–155.
64. Li, N. B., Ren, J., Wang, L., Zhang, G., Hanggi, P., and Li, B. W. (2012) Colloquium: Phononics: Manipulating heat flow with electronic analogs and beyond, *Rev. Mod. Phys.*, **84**, pp. 1045–1066.
65. Tanaka, Y., Yoshida, F., and Tamura, S. (2005) Lattice thermal conductance in nanowires at low temperatures: Breakdown and recovery of quantization, *Phys. Rev. B*, **71**, pp. 205308–205319.
66. Sun, Q. F., Yang, P., and Guo, H. (2002) Four-terminal thermal conductance of mesoscopic sielectric systems, *Phys. Rev. Lett.*, **89**, pp. 175901–175904.
67. Wang, J. and Wang, J. S. (2006) Mode-dependent energy transmission across nanotube junctions calculated with a lattice dynamics approach, *Phys. Rev. B.*, **74**, pp. 054303–054307.

68. Ando, T. (1991) Quantum point contacts in magnetic fields, *Phys. Rev. B*, **44**, pp. 8017–8027.
69. Mingo, N. and Yang, L. (2003) Phonon transport in nanowires coated with an amorphous: an atomistic Green's function approach, *Phys. Rev. B*, **68**, pp. 245406–245417.
70. Mingo, N. (2006) Anharmonic phonon flow through molecular-sized junctions, *Phys. Rev. B*, **74**, pp. 125402–125414.
71. Galperin, M., Nitzan, A. and Ratner, M. A. (2007) Heat conduction in molecular transport junctions, *Phys. Rev. B*, **75**, pp. 155312–155325.
72. Ozpineci, A. and Ciraci, S. (2001) Quantum effects of thermal conductance through atomic chains, *Phys. Rev. B*, **63**, pp. 125415–125419.
73. Wang, J. S., Zeng, N., Wang, J., and Gan, C. K. (2007) Nonequilibrium Green's function method for thermal transport in junctions, *Phys. Rev. E*, **75**, pp. 061128–061142.
74. Xie, Z. X., Li, K. M., Tang, L. M., Pan, C. N. and Chen, K. Q. (2012) Nonlinear phonon transport and ballistic thermal rectification in asymmetric graphene-based three-terminal junctions, *Appl. Phys. Lett.*, **100**, pp. 183110–183113.
75. Xie, Z. X., Tang, L. M., Pan, C. N., Li, K. M., Chen, K. Q., and Duan, W. H. (2012) Enhancement of thermoelectric properties in graphene nanoribbons modulated with stub structures, *Appl. Phys. Lett.*, **100**, pp. 073105–073108.
76. Huang, K. and Zhu, B. (1988) Long-wavelength optic vibrations in a superlattice, *Phys. Rev. B*, **38**, pp. 2183–2186; Dielectric continuum model and Fröhlich interaction in superlattices, *Phys. Rev. B*, **38**, pp. 13377–13386.
77. Sood, A. K., Menendez, J., Cardona, M., and Ploog, K. (1985) Interface vibrational modes in GaAs-AlAs superlattices, *Phys. Rev. Lett.*, **54**, pp. 2115–2118.
78. Hurley, D. C., Tamura, S., Wolfe, J. P., and Morkoc, H. (1987) Imaging of acoustic phonon stop band in superlattices, *Phys. Rev. Lett.*, **58**, pp. 2446–2449.
79. Wang, X. F., Kushwaha, M. S., and Vasilopoulos, P. (2001) Tunability of acoustic spectral gaps and transmission in periodically stubbed waveguides, *Phys. Rev. B*, **65**, pp. 035107–035116.
80. Nishiguchi, N. (1994) Confined and interface acoustic phonons in a quantum wire, *Phys. Rev. B*, **50**, pp. 10970–10980.
81. Graff, K. F. (1975) *Wave Motion in Elastic Solid* (Oxford University press).

82. Xie, Z. X., Chen, K. Q., and Tang, L. M. (2011) Ballistic phonon thermal conductance in a cylindrical semiconductor nanowire modulated with nanocavity, *J. Appl. Phys.*, **110**, pp. 124321–124328.
83. Xu, H. Q. (1994) Scattering-matrix method for ballistic electron transport: theory and an application to quantum antidot arrays, *Phys. Rev. B*, **50**, pp. 8469–8478.
84. David, Y. K. K. and Inkson, J. C. (1988) Matrix method for tunneling in heterostructures: Resonant tunneling in multilayer systems, *Phys. Rev. B*, **38**, pp. 9945–9951.
85. Li, W. X., Chen, K. Q., Duan, W. H., Wu, J., and Gu, B. L. (2003) Phonon transport and thermal conductivity in dielectric quantum wire, *J. Phys. D: Appl. Phys.*, **36**, pp. 3027–3033.
86. Wu, H., Sprung, D. W. L., Martorell, J., and Klarsfeld, S. (1995) Quantum wire with periodic serial structure, *Phys. Rev. B*, **44**, pp. 6351–6360.
87. Valle, M. D., Tejedor, C. and Cuniberti, G. (2005) Defective transport prope ties of three-terminal carbon nanotube junctions, *Phys. Rev. B*, **71**, pp. 125306–125313.
88. Brusheim, P., Csontos, D., Zulicke, U. and Xu, H. Q. (2008) Multiterminal multimode spin-dependent scattering-matrix formalism: Electron and hole quantum spin transport in multiterminal junctions, *Phys. Rev. B*, **78**, pp. 085301–085310.
89. Li, W. X., Chen, K. Q., Duan, W. H., Wu, J., and Gu, B. L. (2004) Acoustic phonon mode splitting behavior of an asymmetric y-branch three-terminal junction, *Appl. Phys. Lett.*, **85**, pp. 822–824.
90. Ming, Y., Wang, Z. X., Li, Q., and Ding, Z. J. (2007) Nonlinear thermal properties of three-terminal mesoscopic dielectric systems, *Appl. Phys. Lett.*, **91**, pp. 143508–143510.
91. Zhang, L., Wang, J. S. and Li, B. W. (2010) Ballistic thermal rectification in nanoscale three-terminal junctions, *Phys. Rev. B*, **81**, pp. 100301–100304.
92. Ming, Y., Wang, Z. X., Ding, Z. J., and Li, H. M. (2010) Ballistic thermal rectification in asymmetric three-terminal mesoscopic dielectric systems, *New Journal of Physics*, **12**, pp. 103041–103050.

Chapter 5

Surface Functionalization–Induced Thermal Conductivity Attenuation in Silicon Nanowires: A Molecular Dynamics Study

Hai-Peng Li[a,b] **and Rui-Qin Zhang**[b]

[a]*Department of Physics, China University of Mining and Technology, Xuzhou, P. R. China*
[b]*Department of Physics and Materials Science, City University of Hong Kong, Hong Kong SAR, P. R. China*
aprqz@cityu.edu.hk

In this chapter, we review our work on the phonon thermal conductivity of hydrogenated and nitrogenated silicon nanowires (SiNWs) at room temperature using equilibrium molecular dynamics (EMD) simulations. Our work focused on a systematical study on the effect of surface passivation on the phonon thermal conductivity of SiNWs. We also provided explanation of the phenomenon derived from the simulations based on the surface atomic structures and phonon spectra of SiNWs considered. Our study is expected to show the importance of surface functionalization in tuning the thermal conductivity of SiNWs.

Nanoscale Energy Transport and Harvesting: A Computational Study
Edited by Gang Zhang

ISBN 978-981-4463-02-7 (Hardcover), 978-981-4463-03-4 (eBook)
www.panstanford.com

5.1 Introduction

Low-dimensional nanostructures have attracted considerable interest in the scientific community due to their dimension-tunable properties, as well as for their potential applications in nanotechnology. Among various low-dimensional structures, one-dimensional SiNWs are especially attractive [1–3] because of their good compatibility with conventional microelectronic technology and their promising candidates as the components of nanoelectronic devices. Although bulk Si is a poor thermoelectric (TE) material, a recent breakthrough made in independent experiments by Hochbaum *et al.* [4] and Bukai *et al.* [5] revealed significantly improved ZT values in SiNWs, which have inspired the research community to reconsider the Si as a TE material.

The performance of TE materials is characterized by the figure of merit, $ZT = \alpha^2\sigma T/\kappa$, where α is the Seebeck coefficient, σ is the electrical conductivity, T is the absolute temperature, and κ is the total thermal conductivity ($\kappa = \kappa_p + \kappa_e$, where κ_p and κ_e are the phonon and electronic contributions, respectively. In semiconductors, the phonon's contribution is dominant at room temperature.) The larger the value of ZT, the better the efficiency of the TE cooler or power generator. The thermal conductivity of semiconductor nanostructures must be minimized to improve the efficiency of TE devices [6]. Therefore, searching for methods that reduce the thermal conductivity of Si nanostructures has become a crucial issue in current Si-based TE research. However, the measurement of the thermal conductivity is very common for bulk materials, but so less for nanostructures that it has been rarely reported in nanostructured materials (e.g., SiNWs [7]). Thus, theoretical modeling and simulation tools are of necessity [8, 9]. Particularly, theoretical model and numerical simulations not only predict the thermal conductivity of nanostructures that have not already been fabricated or that are difficult to characterize in experiment, but also provide information of thermal transport at the nanoscale, which is helpful in designing nanomaterials with tailored thermal conductivity. So far, the thermal conductivity of SiNWs has been extensively studied theoretically by using different methods, including molecular dynamics (MD) simulations [9–15],

Monte Carlo simulations [16], Boltzmann transport equation [17, 18], non-equilibrium Green function approach [19, 20], and first-principles methods [21, 22]. Moreover, reduced thermal conductivity has been observed in Core-Shell nanowires [23], isotope-doping nanowires [24], vacancy-defected nanowires [20], surface-decorated nanowires [25], surface-disordered nanowires [26], and superlatticed nanowires [27, 28]. Such findings provide new strategies to search for high-performance Si-based TE materials by reducing the thermal conductivity.

SiNWs have a large surface-to-volume ratio. At nanometer scale, the surface-to-volume ratio increases with decreasing size, leading to increased significance of the surface in SiNWs. The properties of SiNWs differ remarkably from those of bulk Si due to the surface effect. In addition, the large surface-to-volume ratio of SiNWs provides plenty of room for surface functionalization/modification [29]. SiNWs can be passivated by various radicals, such as hydrogen (H), oxygen, nitrogen (N), phosphorus, fluorine (F) or hydroxyl (OH) groups, which have been observed in experiments [3, 30–33] and studied by computational modeling [29, 34–43]. For example, theoretical study has shown that nitrogen atoms on the surface of Si nanostructures can stabilize the structures and clean the gap states; this is important for visible luminescence in Si nanomaterials [34]. Additionally, first-principles calculations done by us [38] revealed the formation of N–Si–N–Si chains due to nitrogen atoms adsorbed on the (110) facets of [112] SiNWs, which sequentially induces metallization (Fig. 5.1). Their calculations were conducted using density functional theory (DFT) with the generalized gradient approximate (GGA) that is available in the VASP method. The energy cutoff was set at 420 eV, and the GGA was used with the PW91 exchange-correlation functional. The distance between neighboring wires was greater than 0.9 nm to reduce the cell-to-cell interaction, and a $1 \times 1 \times 6$ k-point grid was used for the SiNWs. Recent experimental study demonstrated an effective doping of SiNWs induced by surface passivant [3]. The stability and doping level tunability of SiNWs with alkene adsorption were also observed by us [41], as shown in Fig. 5.2. In their study, DFT calculations were also used within the local density approximation and the projected augmented wave method.

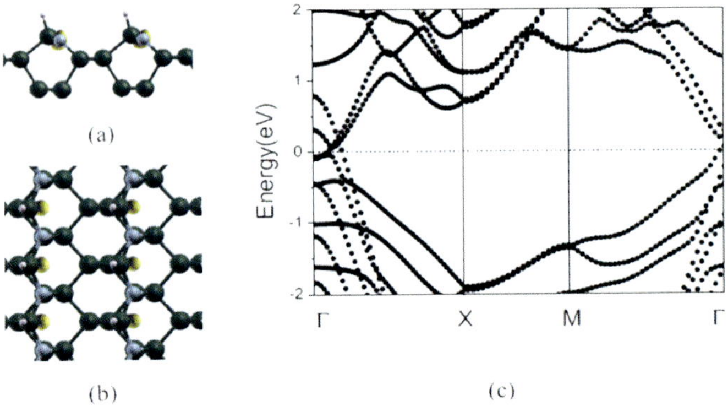

Figure 5.1 (a), (b) the geometry and (c) band structures of silicon thin films with nitrogen adsorption. Here silicon, hydrogen, and nitrogen atoms are displayed by green, white, and blue balls, respectively. Reprinted with permission from [38]. Copyright © 2009, AIP Publishing LLC.

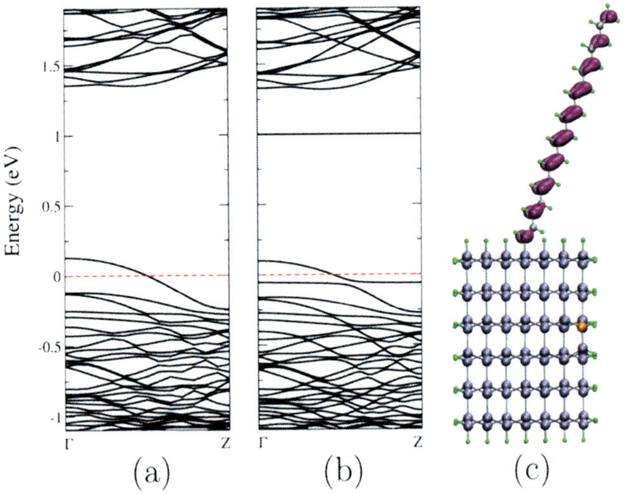

Figure 5.2 Band structures of (a) the B-doped [112] SiNW, (b) the B-doped [112] SiNW with the adsorption of alkenyl chains, and (c) the electronic density from the alkenyl chain. Reprinted with permission from [41]. Copyright © 2011, AIP Publishing LLC.

In addition to the electronic properties, thermal conductivity of SiNWs is also expected to be dependent on surface functionalization because phonon scatterings at the surface can significantly affect the lattice thermal conductivity [44]. Through functionalizing the surface of SiNWs, the thermal conductivity can be significantly modulated [25] and the electronic conductivity can be even increased [3, 41], which is essential for TE application. Therefore, the effect of surface functionalization on the thermal conductivity of SiNWs is attracting great attention in the scientific community. Recent simulations also showed surface modification induced attenuation of the thermal conductivity in carbon nanotubes and diamond nanorods [45, 46]. However, the thermal conductivity of surface-passivated SiNWs has been little studied so far [42, 43].

In this chapter, we review our work on the phonon thermal conductivity of hydrogenated and nitrogenated SiNWs at room temperature using EMD simulations. Our work focused on a systematical study on the effect of surface passivation on the phonon thermal conductivity of SiNWs. We also provided explanation of the phenomenon derived from the simulations based on the surface atomic structures and phonon spectra of SiNWs considered. Our study aims to show the importance of surface functionalization in tuning the thermal conductivity of SiNWs.

5.2 Model and Method

5.2.1 *Structural Model*

SiNWs are oriented along the [001] direction with square cross-section, and the (110)-type surfaces in transverse directions (x and y axes, respectively) and longitudinal direction along the z axis are considered for study. The lattice constants are $a_x = a_y =$ 0.384 nm, and $a_z = 0.543$ nm. The naked surface SiNWs are cut from diamond-structured bulk Si. Then the surface dangling bonds are terminated with hydrogen atoms to construct the hydrogenated SiNWs (H–SiNWs). After that, one Si monohydride (–SiH) is replaced by one nitrogen atom to construct nitrogenated SiNWs (N-SiNWs). Figure 5.3 shows the 2.36 nm^2 cross-section SiNWs passivated by

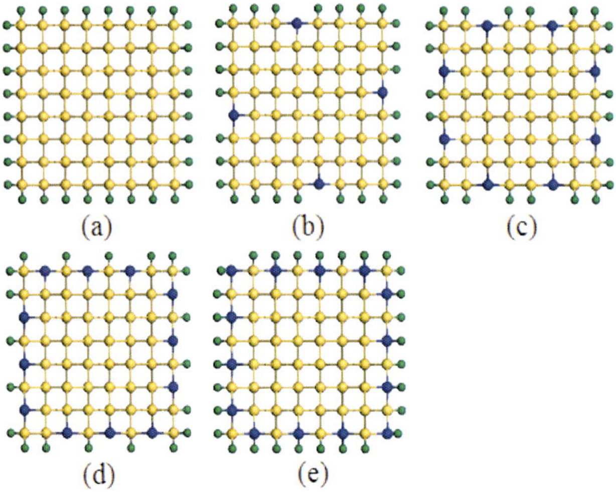

Figure 5.3 Cross-section of SiNWs with different surface nitrogenation ratio: (a) 0%; (b) 14%; (c) 29%; (d) 43%; and (e) 50%. Here silicon, hydrogen, and nitrogen atoms are displayed by yellow, green, and blue balls, respectively. The cross-sectional area is 2.36 nm^2. Reprinted with permission from [43]. Copyright © 2012, City University of Hong Kong.

different numbers of nitrogen atoms. Here surface nitrogenation ratio denotes the ratio of the number of nitrogen atoms in N–SiNW to the number of surface Si atoms in H–SiNW.

5.2.2 *Green–Kubo MD Method*

EMD simulation has emerged as a powerful tool for predicting the lattice thermal conductivity in nanomaterials [10–13]. One major advantage of this method is that it does not rely on too many approximations and assumptions of the detailed phonon processes, which are necessities in solution of Boltzmann transport equation [11]. In MD simulations, interaction between atoms is described by Tersoff potential [47] that changes chemical bond strength according to the local coordination environment [28],

$$V_{ij} = f_C(r_{ij})[f_R(r_{ij}) + B_{ij} f_A(r_{ij})], \tag{5.1}$$

where f_C is a smooth cut-off function, f_A and f_R are attractive and repulsive pair potential functions, and the bond-order term

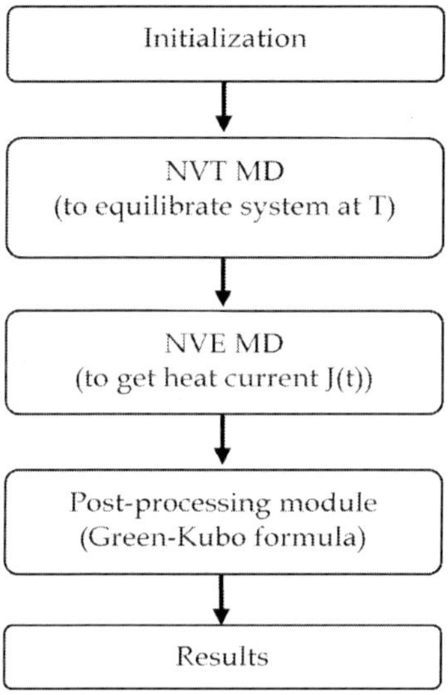

Figure 5.4 Basic algorithm of Green–Kubo MD simulations.

B_{ij} is a function of the number of neighboring atoms, bond angles, and bond lengths. The detailed expressions of Tersoff potential functions and the parameters for silicon, nitrogen, and hydrogen atoms are reported by De Brito Mota *et al.* [48]. Tersoff potential has been successfully applied to calculate thermal conductivity of SiNWs [13, 17] and SiGe NWs [28].

Figure 5.4 shows the basic algorithm of the EMD method. After a NVT (constant particle number N, volume V, and temperature T) equilibration, NVE (constant N, V, and total energy E) MD simulations are carried out to obtain heat current $\vec{J}(t)$ which is given by

$$\vec{J}(t) = \frac{d}{dt} \sum_i (\vec{r}_i E_i), \tag{5.2}$$

where $\vec{r}_i$ and E_i are the time-dependent atomic position vector and the microscopic total energy of atom i, respectively. The thermal conductivity along the z-axial direction is then calculated with the Green–Kubo formula [49],

$$\kappa_z = \frac{1}{V k_B T^2} \int_0^\infty \langle J_z(0) J_z(t) \rangle dt, \tag{5.3}$$

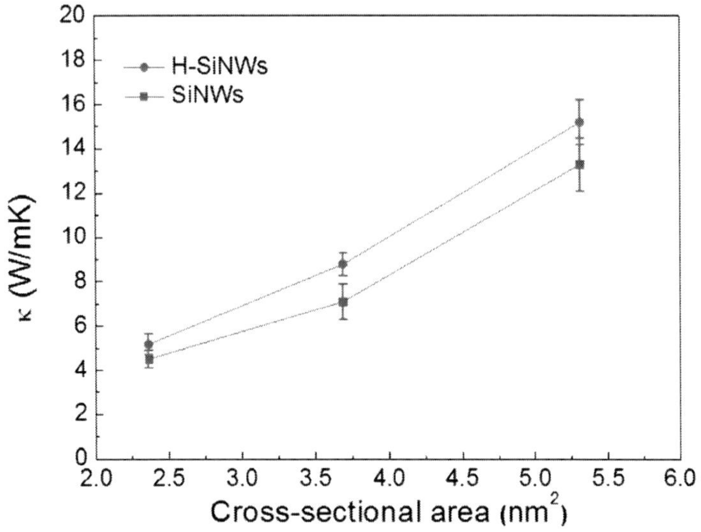

Figure 5.5 Cross-sectional area dependent thermal conductivity for naked SiNWs and H-SiNWs at 300 K (Left to right: 2.36 nm^2, 3.69 nm^2, and 5.31 nm^2). Reprinted with permission from [43]. Copyright © 2012, City University of Hong Kong.

where k_B is the Boltzmann constant, and $J_z(t)$ is the z-component of heat current vector.

5.3 Surface Hydrogenation Effect on the Thermal Conductivity of SiNWs

Simulations are preformed using different lengths of naked SiNWs ranging from 2.72 nm to 8.68 nm, and the results show that the longitudinal thermal conductivity becomes length-independent when the naked SiNWs is longer than 6.52 nm, which corresponds to 12 unit cells. Hydrogenation can induce slight increase in the thermal conductivity, by 15–25% (Fig. 5.5) compared with that of naked surface SiNWs. This phenomenon of hydrogenation induced increase in the thermal conductivity can be correlated with their structures and lattice vibration properties.

The average bond length of diamond-like Si is 0.235 nm. For naked surface SiNWs, our simulations show shortening of Si–Si bonds (by 0.004 nm) in the surface layer compared with the diamond-like case due to surface reconstruction. By contrast, the inner part of the SiNWs retains the diamond lattice. This remarkable surface lattice contraction effect has also been observed for the Si clusters by tight-binding MD (TBMD) simulations [50] and by experiment [51]. Compared to naked SiNWs, passivation by hydrogen saturates the dangling bonds and circumvents the surface reconstruction. For the calculated fully-hydrogenated SiNWs (i.e., H–SiNWs), no lattice contraction is observed while very small lattice expansions (by 0.001 nm) is only found in the surface layer due to hydrogen bond effect, consistent well with TBMD results [52]. Therefore, the bond length of SiNWs in the surface layer is remarkably different between naked surface and H–surface. The bond length of H–surface is higher than that of naked surface, although the inner case is most similar to the bulk Si. Thus the particular surface structures of SiNWs may have an important role in the vibrational modes and thereby the thermal conductivity.

Figure 5.6 shows plots of the phonon density of state (PDOS) spectra for Si atom in the inner and surface layers of naked SiNW and H-SiNW (the cross-sectional area is 2.36 nm^2), respectively. Both 3.69 nm^2 and 5.31 nm^2 SiNWs have similar PDOS spectra. Here PDOS is computed from the Fourier transform of the velocity autocorrelation function in MD simulation [53]. For inner Si atoms, the PDOS curve does not change much with terminated hydrogen layers, as expected from bulk-like behavior. By contrast, for naked SiNWs, the high frequency optical modes of surface Si atoms exhibit a significant blueshift relative to that of inner parts due to the surface lattice contraction effect—a shorter bond length implies a larger force constant and a higher vibration frequency. Compared with PDOS of naked SiNWs, passivation by hydrogen can induce and increase the overlaps of PDOS between surface and inner atoms in H–SiNW, which is benefit to the thermal conductivity. Therefore, the hydrogenation-induced increase in the phonon thermal conductivity is attributed to the specific changes in surface atom vibrational modes.

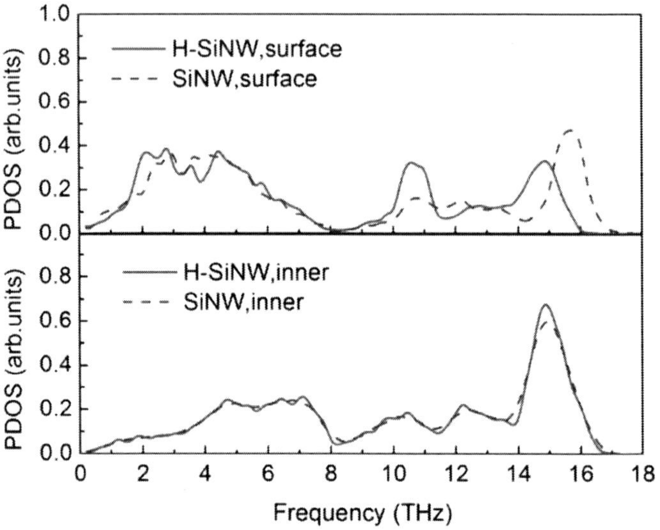

Figure 5.6 PDOS spectra along the longitude direction for Si atoms at the surface layer (top panel) and inner layer (bottom panel) of the naked SiNW (blue dash line) and H-SiNW (red solid line), respectively. The cross-sectional areas of naked SiNW and H-SiNW are 2.36 nm^2. Reprinted with permission from [43]. Copyright © 2012, City University of Hong Kong.

5.4 Surface Nitrogenation Effect on the Thermal Conductivity of SiNWs

Figure 5.7 shows the longitudinal thermal conductivity of nitrogenated SiNWs as a function of surface nitrogenation ratio at 300 K. The nitrogenated SiNWs show remarkably smaller thermal conductivities compared to the fully hydrogenated SiNWs. Nitrogenation of 14% of the surface Si atoms induces the reduction of the thermal conductivity by ~43%, whereas 29% and 43% surface nitrogenation further lower the thermal conductivity by ~60% and ~75%, respectively. The 50% surface nitrogenation has negligible additional effect on the thermal conductivity compared with the case of 43% surface nitrogenation. This nitrogenation-induced attenuation in thermal conductivity has been experimentally confirmed in nitrogenated nanocrystalline diamond (NCD)

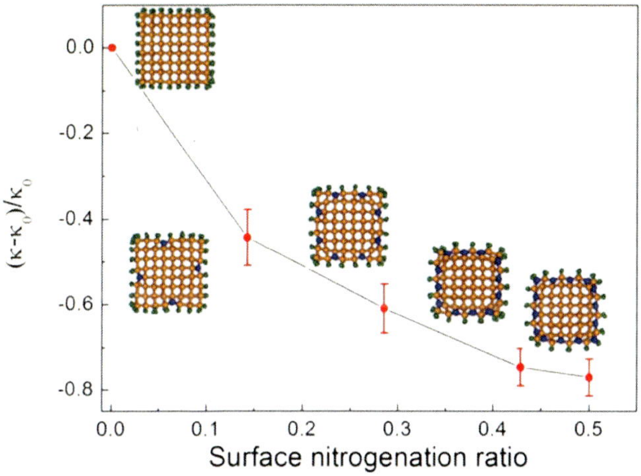

Figure 5.7 Variation of calculated longitudinal thermal conductivity (κ) of nitrogenated SiNWs as a function of surface nitrogenation ratio, compared to thermal conductivity of H-SiNW (κ_0). The cross-sectional area is 2.36 nm^2. Reprinted with permission from [43]. Copyright © 2012, City University of Hong Kong.

films on Si, that is, 25% nitrogenation of an NCD film results in a 50% reduction in thermal conductivity at room temperature [54].

In our simulations, the average N–Si bond length is approximatly 0.176 nm in N–SiNWs, which is slightly larger than the N–Si bond length (0.173 nm [38]) in the $N(SiH_3)_3$ molecule but much smaller than the standard Si–Si bond length (0.235 nm) in bulk silicon. Therefore, surface nitrogenation changes the surface geometry of studied SiNWs and induces surface lattice deformations. Additionally nitrogenation favors the formation of *surface defects* in SiNWs, which leads to enhanced thermal diffuse scattering [55]. The defects near the surface with higher concentrations will scatter phonons more effectively when the surface nitrogenation is increased, thereby further decreasing the thermal conductivity of the SiNWs. In semiconductor nanowires, charge carriers are naturally pushed away from the surface [44]. Thus, surface scattering by nitrogenation can reduce the thermal conductivity at a higher degree than the electrical conductivity, which is beneficial to obtaining higher TE performance [4].

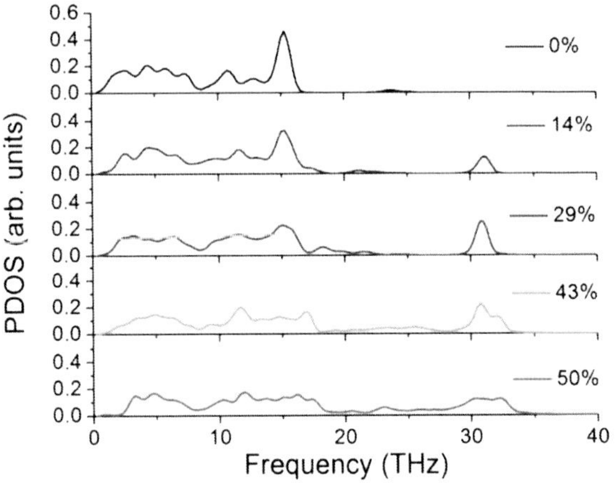

Figure 5.8 PDOS spectra along the longitudinal direction for Si and nitrogen atoms in the 2.36 nm^2 cross-section SiNWs with different degrees of surface nitrogenation. The curves obtained from MD simulations are drawn by a Gaussian smoothing with a width of 0.75 THz. Reprinted with permission from [43]. Copyright © 2012, City University of Hong Kong.

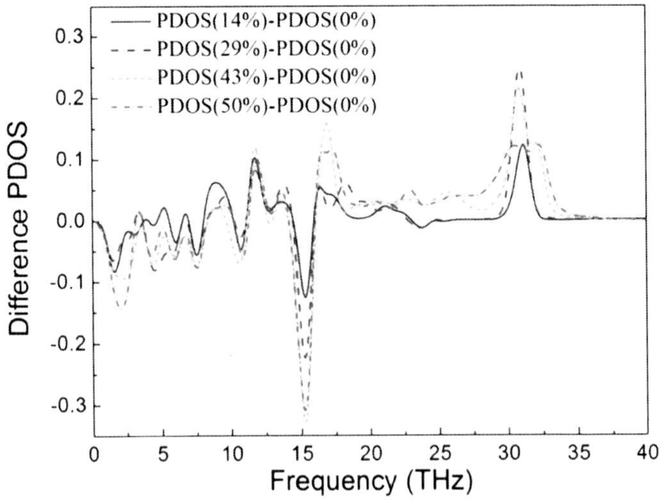

Figure 5.9 Difference PDOS spectra for silicon and nitrogen atoms in the 2.36 nm^2 cross-section SiNWs with surface nitrogenation ratio 14%, 29%, 43%, and 50%. The reference spectrum is the PDOS of H–SiNWs (i.e., surface nitrogenation ratio = 0%). The curves obtained from MD simulations are drawn by a Gaussian smoothing with a width of 0.75 THz. Reprinted with permission from [43]. Copyright © 2012, City University of Hong Kong.

Figure 5.8 plots the PDOS spectra for Si and nitrogen atoms in N–SiNWs. Given the surface lattice deformation, the PDOS spectra of N–SiNWs are quite different from that of H–SiNW (i.e., 0% surface nitrogenation). With the degree of nitrogenation increasing, the Si–Si vibrational peak near 16 THz is depressed and blue shifted whereas a high frequency peak near 30 THz, which corresponds to Si–N vibrational mode (830 cm^{-1} – 890 cm^{-1} [56]), appears. In addition, when the surface nitrogenation ratio increases from 0% to 50%, the PDOS from Si atoms and nitrogen atoms show a depression in the low-frequency part (Fig. 5.9), resulting from the lattice mismatch and difference in the atomic mass between the Si core and N-ad-layer. The remarkable reduction in the low-frequency PDOS mainly cause the decrease in the thermal conductivity for N-SiNWs. Therefore, the surface nitrogenation induced attenuation in thermal conductivity is attributed to the remarkable phonon scattering by surface defects [46] (the presence of lattice mismatch and passivated atoms near the surface), which dampen some of the lattice vibrational modes [23].

5.5 Conclusions and Remarks

In summary, the phonon thermal conductivities of surface hydrogenated and nitrogenated thin SiNWs investigated using equilibrium MD simulations, respectively, have been reviewed. Surface hydrogen passivation can increase the thermal conductivity compared to that of naked SiNWs without surface passivation. Hydrogenation can saturate the dangling bonds and reduce the lattice mismatch between the inner part and the surface of SiNWs, and hence increase the overlaps of the phonon spectra between surface and inner atoms. Such specific changes in the lattice structures and vibrational modes of surface Si atoms cause the increase of thermal conductivity in hydrogenated SiNWs. In addition, surface nitrogen passivation can significantly reduce the thermal conductivity of SiNWs. In Particular, 50% surface nitrogenation on SiNWs can lower the thermal conductivity by about 75% with respect to that of fully hydrogenated SiNWs. This reduction in the thermal conductivity arises from the suppression of some lattice vibrational modes mainly because of the significant effect of phonon-surface defect scattering. Our study highlights the importance of surface modification in the thermal conductivity of SiNWs. The surface atoms have a much more dominant effect on the overall thermal-transfer properties because of the large surface-to-volume ratio, especially for thin SiNWs. Surface modification can become an effective method of tuning the thermal conductivity of thin SiNWs. This finding will be of great interest to both scientific researchers and technologists, as they can serve as highly useful references for Si-based TE applications.

References

1. Cui, Y., and Lieber, C. M., Jr. (2001) Functional nanoscale electronic devices assembled using silicon nanowire building blocks, *Science*, **291**(5505), 851–853.
2. Gudiksen, M. S., Lauhon, L. J., Wang, J., Smith, D. C., and Lieber, C. M., Jr. (2002) Growth of nanowire superlattice structures for nanoscale photonics and electronics, *Nature*, **415**, 617–620.
3. Guo, C. S., Luo, L. B., Yuan, G. D., Yang, X. B., Zhang, R. Q., Zhang, W. J., and Lee, S. T., Jr. (2009) Surface passivation and transfer doping of silicon nanowires, *Angew. Chem. Int. Ed.*, **48**(52), 9896–9900.
4. Hochbaum, A. I., Chen, R., Delgado, R. D., Liang, W., Garnett, E. C., Najarian, M., Majumdar, A., and Yang, P., Jr. (2008) Enhanced thermoelectric performance of rough silicon nanowires, *Nature*, **451**(7175), 163–167.
5. Boukai, A. I., Bunimovich, Y., Tahir-Kheli, J., Yu, J.-K., Goddard III, W. A., and Heath, J. R., Jr. (2008) Silicon nanowires as efficient thermoelectric materials, *Nature*, **451**(7175), 168–171.
6. Hicks, L. D., and Dresselhaus, M. S., Jr. (1993) Thermoelectric figure of merit of a one-dimensional conductor, *Phys. Rev. B*, **47**(24), 16631–16634.
7. Li, D., Wu, Y., Kim, P., Shi, L., Yang, P., and Maujmdar A., Jr. (2003) Thermal conductivity of individual silicon nanowires, *Appl. Phys. Lett.*, **83**(14), 2934.
8. Chantrenne, P., Barrat, J. L., Blase, X., and Gale, J. D., Jr. (2005) An analytical model for the thermal conductivity of silicon nanostructures, *J. Appl. Phys.*, **97**(10), 104318.

9. Abs da Cruz, C., Termentzidis, K., Chantrenne, P., and Kleber, X., Jr. (2011) Molecular dynamics simulations for the prediction of thermal conductivity of bulk silicon and silicon nanowires: Influence of interatomic potentials and boundary conditions, *J. Appl. Phys.*, **110**(3) 034309.
10. Volz, S. G., and Chen, G., Jr. (1999) Molecular dynamics simulation of thermal conductivity of silicon nanowires, *Appl. Phys. Lett.*, **75**(14), 2056.
11. Schelling, P. K., Phillpot, S. R., and Keblinski, P., Jr. (2002) Comparison of atomic-level simulation methods for computing thermal conductivity, *Phys. Rev. B*, **65**(14), 144306.
12. Chen, J., Zhang, G., and Li, B., Jr. (2010) Remarkable reduction of thermal conductivity in silicon nanotubes, *Nano Lett.*, **10**(10), 3978–3983.
13. Li, X., Maute, K., Dunn, M. L., and Yang, R., Jr. (2010) Strain effects on the thermal conductivity of nanostructures, *Phys. Rev. B*, **81**(24), 245318.
14. Sellan, D. P., Landry, E. S., Turney, J. E., McGaughey, A. J. H., and Amon C. H., Jr. (2010) Size effects in molecular dynamics thermal conductivity predictions, *Phys. Rev. B*, **81**(21), 214305.
15. Wang, S. C., Liang, X. G., Xu, X. H., and Ohara, T., Jr. (2009) Thermal conductivity of silicon nanowire by nonequilibrium molecular dynamics simulations, *J. Appl. Phys.*, **105**(1), 014316.
16. Chen, Y., Li, D., Lukes, J. R., and Majumdar, A., Jr. (2005) Monte Carlo simulation of silicon nanowire thermal conductivity, *J. Heat Transfer*, **127**(10), 1129–1137.
17. Donadio, D., and Galli, G., Jr. (2010) Temperature dependence of the thermal conductivity of thin silicon nanowires, *Nano Lett.*, **10**(3), 847–851.
18. Mingo, N., Jr. (2003) Calculation of Si nanowire thermal conductivity using complete phonon dispersion relations, *Phys. Rev. B*, **68**(11), 113308.
19. Markussen, T., Jauho, A.-P., and Brandbyge, M., Jr. (2009) Electron and phonon transport in silicon nanowires: atomistic approach to thermoelectric properties, *Phys. Rev. B*, **79**(3), 035415.
20. Yamamoto, K., Ishii, H., Kobayashi, N., and Hirose, K., Jr. (2011) Effects of vacancy defects on thermal conduction of silicon nanowire: Nonequilibrium Green's function approach, *Appl. Phys. Express*, **4**, 085001.
21. Esfarjani, K., Chen, G., and Stokes, H. T., Jr. (2011) Heat transport in silicon from first-principles calculations, *Phys. Rev. B*, **84**(8), 085204.

22. Broido, D. A., Malorny, M., Birner, G., Mingo, N., and Stewart, D. A., Jr. (2007) Intrinsic lattice thermal conductivity of semiconductors from first principles, *Appl. Phys. Lett.*, **91**(23), 231922.
23. Hu, M. Giapis, K. P., Goicochea, J. V., Zhang, X., and Poulikakos, D., Jr. (2011) Significant reduction of thermal conductivity in Si/Ge Core-Shell nanowire, *Nano Lett.*, **11**(2), 618–623.
24. Yang, N., Zhang, G., and Li, B., Jr. (2008) Ultralow thermal conductivity of isotope-doped silicon nanowires, *Nano Lett.*, **8**(1), 276–280.
25. Markussen, T., Jauho, A., and Brandbyge, M., Jr. (2009) Surfacedecorated silicon nanowires: a route to high-ZT thermoelectric, *Phys. Rev. Lett.*, **103**(5), 055502.
26. Donadio, D., and Galli, G., Jr. (2009) Atomistic simulations of heat transport in silicon nanowires, *Phys. Rev. Lett.*, **102**(19), 195901.
27. Shelley, M., and Mostofi, A. A., Jr. (2011) Prediction of high zT in thermo-electric silicon nanowires with axial germanium heterostructures, *EPL*, **94**(6), 67001.
28. Chan, M. K. Y., Reed, J., Donadio, D., Mueller, T., Meng, Y. S., Galli, G., and Ceder, G., Jr. (2010) Cluster expansion and optimization of thermal conductivity in SiGe nanowires, *Phys. Rev. B*, **81**(17), 174303.
29. Zhang, R. Q., Lifshitz, Y., Ma, D. D. D., Zhao, Y. L., Frauenheim, T., Lee, S. T., and Tong, S. Y., Jr. (2005) Structures and energetics of hydrogen-terminated silicon nanowire surfaces, *J. Chem. Phys.*, **123**(14), 144703.
30. Lin, Y., Gonzalez Szwacki, N., and Yakobson B. I., J. (2007) Quasi-one-dimensional silicon nanostructures, in *Nanosilicon* ed. Kumar V. (Elsevier pp. 289–313).
31. Zhang, R. Q., Lifshitz, Y., and Lee, S. T., Jr. (2003) Oxide-assisted growth of semiconducting nanowires, *Adv. Mater.*, **15**(7–8), 635–640.
32. Dai, M., Wang, Y., Kwon, J., Halls, M. D., and Chabal, Y. J., Jr. (2009) Nitrogen interaction with hydrogen-terminated silicon surfaces, *Nature Mater.*, **8**, 825–830.
33. Wang, N., Cai, Y., and Zhang, R. Q., Jr. (2008) Growth of nanowires, *Mater. Sci. Eng. R*, **60**(1–6), 1–51.
34. Zhang, R. Q., Costa, J., and Bertran, E., Jr. (1996) Role of structural saturation and geometry in the luminescence of silicon-based nanostructured materials, *Phys. Rev. B*, **53**(14), 7847–7850.
35. Lu, A. J., Zhang, R. Q., and Lee, S. T., Jr. (2007) Tunable electronic band structures of hydrogen-terminated [112] silicon nanowires, *Appl. Phys. Lett.*, **92**(20), 203109.

36. Lu, A. J., Zhang, R. Q., and Lee, S. T., Jr. (2008) Unique electronic band structures of hydrogen-terminated [112] silicon nanowires, *Nanotechnology*, **19**(3), 035708.

37. Yang, X. B., Guo, C. S., and Zhang, R. Q., Jr. (2009) Prediction of surface passivation doping of silicon nanowires with phosphoru, *Appl. Phys. Lett.*, **95**(19), 193105.

38. Yang, X. B., and Zhang, R. Q., Jr. (2009) Metallization induced by nitrogen atom adsorption on silicon nanofilms and nanowires, *Appl. Phys. Lett.*, **94**(11), 113101.

39. Xu, H., Yang, X. B., Guo, C. S., and Zhang, R. Q., Jr. (2009) An energetic stability predictor of hydrogen-terminated Si nanostructures, *Appl. Phys. Lett.*, **95**(25), 253106.

40. Xu, H., Zhang, X. H., and Zhang, R. Q., Jr. (2009) First-principles study of silicon bulk and nanowire (111) surfaces terminated with trihydrides: Symmetric, rotated, and tilted, *Phys. Rev. B*, **80**(7), 075428.

41. Xu, H., Yang, X. B., Zhang, C., Lu, A. J., and Zhang, R. Q., Jr. (2011) Stabilizing and activating dopants in <112> silicon nanowires by alkene adsorptions: a first-principles study, *Appl. Phys. Lett.*, **98**(7), 073115.

42. Li, H. P., De Sarkar, A., and Zhang R. Q., Jr. (2011) Surface-nitrogenation–induced thermal conductivity attenuation in silicon nanowires, *EPL*, **96**(5), 56007.

43. Li, H. P., PhD thesis (2012) Molecular dynamics simulations of phonon thermal transport in low-dimensional silicon structures, City University of Hong Kong, pp. 85–102.

44. Martin, P., Aksamija, Z., Pop, E., and Ravaioli, U., Jr. (2010) Reduced thermal conductivity in nanoengineered rough Ge and GaAs nanowires, *Nano Lett.*, **10**(4), 1120–1124.

45. Padgett, C. W., and Brenner, D. W., Jr. (2004) Influence of chemisorption on the thermal conductivity of single-wall carbon nanotubes, *Nano Lett.*, **4**(6), 1051–1053.

46. Padgett, C. W., Shenderova, O., and Brenner, D. W., Jr. (2006) Thermal conductivity of diamond nanorods: Molecular simulation and scaling relations, *Nano Lett.*, **6**(8), 1827–1831.

47. Tersoff, J., Jr. (1989) Modeling solid-state chemistry: Interatomic potentials for multicomponent systems, *Phys. Rev. B*, **39**(8), 5566–5568.

48. De Brito Mota, F., Justo, J. F., and Fazzio, A. Jr. (1999) Hydrogen role on the properties of amorphous silicon nitride, *J. Appl. Phys.*, **86**(4), 1843–1847.

49. Kubo, R., Jr. (1966) The fluctuation–dissipation theorem, *Rep. Prog. Phys.*, **29**, 255–284.
50. Yu, D. K., Zhang, R. Q., and Lee, S. T., Jr. (2002) Structural transition in nanosized silicon clusters, *Phys. Rev. B*, **65**(24), 245417.
51. Hofmeister, H., Huisken, F., and Kohn, B., Jr. (1999) Lattice contraction in nanosized silicon particles produced by laser pyrolysis of silane, *Eur. Phys. J. D*, **9**(1), 137–140.
52. Yu, D. K., Zhang, R. Q., and Lee, S. T., Jr. (2002) Structural properties of hydrogenated silicon nanocrystals and nanoclusters, *J. Appl. Phys.*, **92**(12), 7453.
53. Dickey, J. M., and Paskin, A., Jr. (1969) Computer simulation of the lattice dynamics of solids, *Phys. Rev.*, **188**(3), 1407–1418.
54. Shamsa, M., Ghosh, S., Calizo, I., Ralchenko, V., Popovich, A., and Balandin, A. A., Jr. (2008) Thermal conductivity of nitrogenated ultrananocrystalline diamond films on silicon, *J. Appl. Phys.*, **103**(8), 083538.
55. Yang, H. N., and Lu, T. M., Jr. (1991) Enhancement of thermal diffuse scattering by surface defects, *Phys. Rev. B*, **44**(20), 11457–11464.
56. Tsu, D. V., Lucovsky, G., and Mantini, M. J., Jr. (1986) Local atomic structure in thin films of silicon nitride and silicon dimide produced by remote plasma-enhanced chemical-vapor deposition, *Phys. Rev. B*, **33**(10), 7069–7076.

Index